Systems Engineering in Agile Software Development

Dallas Rosson

Systems Engineering in Agile Software Development

Springer

Dallas Rosson
Industrial and Systems Engineering
University of Washington
Seattle, WA, USA

ISBN 978-3-032-14612-0 ISBN 978-3-032-14613-7 (eBook)
https://doi.org/10.1007/978-3-032-14613-7

This Springer imprint is published by the registered company Springer Nature Switzerland AG
The registered company address is: Gewerbestrasse 11, 6330 Cham, Switzerland

I would like to dedicate this textbook to my grandmother, Dorothy Stark. It was her encouragement early in life that set me on the path to earning a doctorate. I am sure she is smiling down on me from above for not only achieving that dream but always striving to continue being a lifelong learner.

Back of Cover Description

Many software projects lack the benefits of established systems engineering methodologies and good engineering rigor, failing to meet customer needs and expectations. Software projects that are in critical systems, such as those in the financial, medical, aviation, and defense industries cannot afford failures or major defect introduction into their systems as they can lead to dire outcomes for individual and corporate safety and welfare. Software developers trained in the use of the various Agile frameworks are frequently strongly opposed to any development methodology that could be viewed as infringing on the principles of the Agile Manifesto. Agile projects, by their nature, embrace the concept of change, but uncontrolled change leads to project failures and the introduction of defects, whereas controlled change can lead to sustained and innovative forward progress. Through the thoughtful implementation of systems engineering methodologies, such as requirements management, systems modeling, and rigorous quality assurance, project risks associated with cost, schedule, and performance are greatly diminished as well as the realization of increased system stability. This text details the struggles of Agile methodologies in meeting the needs of critical software systems development and presents a methodology to bridge the gap between traditional engineering methods, such as Waterfall or Predictive development, and the industry standard of Agile development.

Acknowledgements

I am indebted to numerous mentors, colleagues, and most of all, my family, for support over the many years I spent pursuing my passion for education and discovering my passion for systems engineering and software development, leading to this book. Of note, the mentorship of Dr. Thomas Bradley and Dr. Juming Tang has been instrumental in guiding me along my path. I would be remiss to not acknowledge my parents, my spouse, and my children, who have dealt with years of listening to me prattle about a topic which, I am sure, is not as interesting to them as it is to me and for the encouragement to pursue my dreams of publishing this text. Finally, I would like to give special thanks to Dr. Daniel Herber who offered editorial insights and whose published works are key to some of the Model Based Systems Engineering methods implemented throughout the book.

Competing Interests The author has no competing interests to declare that are relevant to the content of this manuscript.

Contents

About the Author

Dallas Rosson launched his government service career in the U.S. Navy as an Electronic Technician before advancing to Radioman, specializing in communications and electronic warfare. He concluded his Naval service in 2010 as a First-Class Petty Officer, having completed six ballistic missile deterrent patrols and earned the Submarine Warfare Qualification.

Following his military tenure, Dr. Rosson pursued academic advancement at the University of Washington, earning a Bachelor of Science in Computer Science and Systems. During his studies, he interned as a software developer for the Advanced Skills Management System (ASM) at the Naval Undersea Warfare Center (NUWC) Keyport, WA. Upon graduation, he assumed the role of Network Manager for the North Kitsap School District in Poulsbo, WA. After two years, he returned to federal employment as a computer scientist, conducting software engineering on the Obsolescence Management Information System (OMIS) at NUWC Keyport. Over his tenure at NUWC Keyport, Dr. Rosson has held positions as Software Engineer, Project Manager, Systems Engineer, Principal Engineer, and Chief Engineer, contributing to more than a dozen critical defense software systems, including enterprise-scale deployments utilized across the Department of Defense.

In 2024, Dr. Rosson joined the University of Washington's Industrial and Systems Engineering Department faculty under Professor Juming Tang's supervision, where he instructs courses in systems engineering, including Model-Based Systems Engineering, alongside offering workshops and masterclasses.

Dr. Rosson's educational background is comprehensive, comprising a Bachelor of Science and Master of Science in Computer Science and Systems from the University of Washington, a Master of Science in Systems Engineering from George Washington University, and a Doctor of Engineering in Systems Engineering from Colorado State University, advised by Dr. Thomas Bradley and Ms. Ann Batchelor.

He holds multiple professional certifications, has published scholarly and trade articles, and has been recognized as a keynote speaker.

Education: Doctor of Engineering, Systems Engineering, Colorado State University, Master of Science, Systems Engineering, George Washington University, Master of Science, Computer Science and Systems, University of Washington, Bachelor of Science, Computer Science and Systems, University of Washington.

Professional Certifications: Certified Systems Engineering Professional (CSEP), International Council on Systems Engineering (INCOSE), Project Management Professional (PMP), Project Management Institute, Advanced Certified Scrum Product Owner (ACSPO), Scrum Alliance, Advanced Certified Scrum Master (ACSM), Scrum Alliance, Defense Acquisition Workforce Improvement Act (DAWIA) Engineering Level 2 Certification, Defense Acquisition University, Defense Acquisition Workforce Improvement Act (DAWIA) Information Technology Level 2 Certification, Defense Acquisition University, Defense Acquisition Workforce Improvement Act (DAWIA) Engineering & Technology Management Practitioner Level Certification, Defense Acquisition University, U.S. Navy Cybersecurity Workforce Certification, Software Developer (Specialty 621) Level 3 (Advanced).

Abbreviations

AAF	Adaptive Acquisition Framework
AQDR	Additional Qualification Designator Report
CDR	Critical Design Review
CI/CD	Continuous Integration / Continuous Development
CMMI	Capability Maturity Model Integration
CMP	Configuration Management Plan
CONUS	Inside the 48 Contiguous United States and the District of Columbia
COTS	Commercial Off the Shelf
CRRB	Change Request Review Board
CV	US Navy Abbreviation for Aircraft Carrier
DCS	DoD Control System
DE	Domain Engineering
Defense	Navy & Marine Corps Intranet
DESS	DoD Data Exchange Software System
DLMS	Defense Logistics Management System
DMSMS	Diminishing Manufacturing Sources and Material Shortages
DoD	Department of Defense
DoDAF	Department of Defense Architectural Framework
DPAS	DoD Process Automation System
DQMS	DoD Qualification Management System
DSHS	DoD Software Hosting System
DTSS	DoD Tactical Support System
DWMS	DoD Work Management System
ERB	Enhancement Review Board
EVM	Earned Value Management
FCA	Functional Configuration Audit
FRP	Full Rate Production
IDE	Integrated Development Environment

IETMS	Interactive Electronic Technical Manuals
IMA	Intermediate Maintenance Activity
IMF	Intermediate Maintenance Facility
In	Iteration Number
IPT	Integrated Product Team
LRIP	Low-Rate Initial Production
MBSE	Model Based Systems Engineering
MBSR	Model-based Structured Requirements
MDA	Milestone Decision Authority
MDE	Model Driven Engineering
MVCR	Minimum Viable Capability Release
MVP	Minimum Viable Product
NSWC	Naval Surface Warfare Center
NUWC	Naval Undersea Warfare Center
OBE	Overcome by Events
OCONUS	Outside the Contiguous United States
OMIS	Obsolescence Management Information System
OUSD	Office of the Undersecretary of Defense
PCA	Physical Configuration Audit
PDR	Preliminary Design Review
PPR	Project Performance Review
PRR	Production Readiness Review
PSNS	Puget Sound Naval Shipyard
QCL	Qualification, Certification, Licenses
R4J	Requirements for JIRA
RBR	Release Backlog Review
RMB	Requirements Management Board
Rn	Release Number
RPM	Registered Publication Memorandums
SAFe	Scaled Agile Framework
SETR	Systems Engineering Technical Reviews
SOS	System of Systems
SQL	Structured Query Language
SRB	Stakeholder Review Board
SRD	Software Requirements Document
SRR	Software Requirements Review
SRS	Software Requirements Specification
SRT	Submarine Readiness Tracker
SSBN	US Navy Abbreviation for Sub-surface Ballistic Nuclear Submarine
SSN	US Navy Abbreviation for Sub-surface Nuclear Submarine
T&E	Test & Evaluation
TMIS	Torpedo Management Information System
TRR	Test Readiness Review
TSU	Trainer Scheduling Unit

TYCOM	Type Commander
UDT	User Design Team
UI	User Interface
UML	Unified Modeling Language
USMD	United States Marine Corps
USN	United States Navy
V&V	Verification and Validation

Chapter 1
Introduction

1.1 Why this Book Exists

Software projects in critical domains cannot afford to fail due to concerns with safety, finances, and other criticality drivers and yet, many do. In the context of this text, we shall refer to software projects in the domains of defense, healthcare, finance, and aviation as critical systems or systems in which, at best, failures result in financial harm or, at worst, loss of life. Whether the failure stems from scope creep, misunderstood requirements, or an inability to adapt to changing stakeholder needs, the root cause is often a breakdown in engineering rigor, over reliance on Agile flexibility, or both.

This book introduces a merger of Agile software engineering and traditional Systems Engineering methods to bridge that gap. It presents a practical framework that merges the rigor of systems engineering with the adaptability of Agile Scrum, the Systems Engineering Focused Agile Development (SEFAD) method. Grounded in real-world case studies, validated by simulation and experience, and aligned with modern acquisition policy, this book equips readers to build systems that are both engineered and agile.

1.1.1 Story Time

Before I jump into the topic of merging Agile Scrum with Systems Engineering rigor, I want to share with you a short story. This story revolves around a project that I was part of when I first started working as a civilian federal employee software developer for a logistics management software system in the US Navy. This project had the goal of transitioning a macro-enabled Excel spreadsheet into a working module for the software my team was responsible for. The development was funded

D. Rosson, *Systems Engineering in Agile Software Development*,
https://doi.org/10.1007/978-3-032-14613-7_1

and pushed by mid-level management and did not have full buy-in from the deck plate logistics workers. We were handed the spreadsheet and the macros, in this case, a Visual Basic backend which none of the developers knew the language of, and the understanding that we would have to implement a Poisson distribution as an algorithmic method, which, without getting into too deep of detail, is a discrete probability calculation that isn't very easy to implement in code, especially when one is new to logistics. "Requirements gathering" was spending an hour in a room listening to management speak about what they wanted, a few grunts from the logistics workers, and a smile from my Project Lead. We spent the next 5 months developing a system module based upon the information from that meeting and the spreadsheet. We delivered it, amazingly, on time and within budget. We won awards for the effort and were lauded for our ingenuity.

Fast forward to a few years later. I am the newly minted Project Lead and Systems Engineer over this project. One of my goals was to minimize maintaining modules that weren't often used. I set my IT team into digging into the logs to find out logins, usage, and other metrics to identify our core functionality usage. It doesn't take long for me to get a message over Teams. "Dallas, you need to come look at these stats. You aren't going to believe it." I am sure you all can see where this is going. The module we built those years ago, the very one we had been given awards for creating and were told was going to change how the logisticians work showed usage numbers of nothing. In the previous years there had been a total of five logged users out of 300 logisticians. Of those 5 users, it appeared they simply logged in, looked at it, and never tried again. We had spent hundreds of thousands of dollars developing a system that workers didn't want to meet demands of management. This wasn't the last time that this would occur, but it was the one that planted the seed that would evolve into the topic of this book.

1.2 The Core Challenge

Traditional systems engineering provides a structured approach to managing complexity, utilizing tools such as comprehensive requirements management, system-level design, and rigorous verification practices. These methods enable high-confidence decision-making and traceability throughout the product lifecycle where there is minor risk of scope change and requirements can be elucidated upfront. When used in fast-paced development environments, especially those with changing requirements or where the problem domain is not fully defined, such approaches are often seen as less adaptable to change. The most common of these methods is the Waterfall method.

Agile Scrum has a different structure compared to predictive approaches like Waterfall. Agile approaches aim to incorporate rapid feedback, ongoing user participation, and changing stakeholder requirements, facilitating a flexible and iterative development process that can correspond with customer needs. However, Agile methods may not always provide the structure necessary for maintaining traceable requirements, conducting formal technical reviews, or establishing consistent architectural frameworks, which are often important for mission-critical and safety-critical systems.

The Systems Engineering Focused Agile Development (SEFAD) methodology addresses this challenge by integrating the disciplined approach of systems engineering with the adaptability and flexibility of Agile Scrum, implementing a flexible—rigid—flexible software development execution method. The SEFAD method empowers engineering teams to maintain the ability to respond effectively to evolving stakeholder requirements, as a project team that was utilizing traditional Agile Scrum would, while upholding the engineering rigor, technical rigor, and accountability essential for complex systems and programs, as one would assume with the Waterfall method. This methodology fosters an environment that is receptive and swiftly reactive to change while ensuring engineering and technical rigor is maintained.

1.3 The SEFAD Approach

SEFAD presents a solution to our dilemma using a modified V-Model backbone to establish systems architecture, mission requirements, and design reviews up front, transitioning through a rigid—flexible—rigid execution, and integrating in iterative Agile development.

What does this mean?

- Requirements are not rigid. They are flexible until minimum viable product, or product increment, baseline.
- Mission threads are not ignored. They are mapped, validated, and documented.
- Sprints are not undirected. They are driven by validated models and traced to mission threads and objectives.

The SEFAD model structures software development into four phases:

1. Project definition phase: The development team and/or systems engineers perform needs analysis and requirements management. Key reviews such as the stakeholder needs review (SNR) and the system requirements review (SRR).
2. Modeling and test planning phase: The development team and/or systems engineers perform initial modeling, design verification, test planning, and create the proposed development schedule. The key review for this phase is the Initial Design Review (IDR).
3. Development phase: Agile scrum teams iterate through sprints, delivering usable increments, collaborating directly with stakeholders, and continuously refining backlog items. This phase preserves flexibility while maintaining technical accountability through ongoing Sprint reviews and integration with system models and traceable requirements. Key reviews for this phase are the publish authorization review, sprint review, and final design review (FDR).

4. Project finalization phase: Once the system approaches stability, a return to rigor ensures readiness for production, normally achieved between 80% and 95% project completion based upon stakeholder and Product Owner agreement. The key reviews of the Product Readiness Review (PRR) confirm system completeness, compliance, and documentation. Change control and configuration management become paramount. The project closeout event signals the full completion of the project.

This balance—early structure, midstream agility, and end-stage rigor—is essential for software that must not only work but endure.

1.4 Benefits of the Integrated Approach

Data from simulated and real Agile programs within defense and industrial sectors demonstrates that integrating systems engineering methods into Agile Scrum results in:

- Improved stakeholder communication through traceable needs and requirements.
- Faster adaptation to changing goals without losing architectural stability.
- Greater understanding of system context via modeling and technical documentation.
- Reduced project risk, particularly for cost, schedule, and scope.
- Higher team and customer satisfaction due to aligned expectations and deliverables.
- These results are not simply theory but are drawn from multiple successful project executions.

1.5 What You Will Learn

This text will guide you through:

- A comparison of Agile Scrum and Systems Engineering principles.
- How to perform requirements management and system modeling in an Agile context.
- The SEFAD process, roles, and artifacts.
- How to tailor technical reviews such as SRR, PDR, and CDR into Agile increments.
- Methods for integrating verification and validation into CI/CD pipelines.
- How to apply this methodology in real-world domains like defense logistics, health informatics, and enterprise software systems.
- Each chapter includes real-world case studies, practical exercises, and suggestions for further study.

1.6 Who This Book Is For

This textbook is written for:

- Students studying systems engineering or software engineering.
- Practicing engineers and technical leaders in high-assurance or regulated domains.
- DoD contractors and acquisition professionals needing to align Agile with systems engineering mandates.
- Scrum practitioners who want to introduce structure and traceability into their projects.
- Instructors that are seeking a curriculum-ready text for courses on hybrid development methods.

1.7 The Path Ahead

Agility and engineering rigor do not have to be viewed as enemies. In fact, it is possible to develop in an Agile way while maintaining necessary engineering and technical rigor. Without engineering rigor and structure, Agile descends into chaos; customers and developers can struggle with communicating, requirements are often poorly detailed, and whole project's scopes change. Without Agile, engineering calcifies into irrelevance as it is impossible in modern software engineering to develop a system with requirements and the problem domain fully known upfront. The most successful software development teams know how to balance the art of adaptation with the rigor of systems engineering. The goal of this text is to impart that mastery upon you, the reader.

Chapter 2
Foundations of Systems Engineering and Agile Scrum

2.1 Systems Engineering in Software Development

Systems engineering is essential to software development in critical systems. Though it is essential, it is often overlooked for the need to "move fast and break things" that most Agile developers desire. There is a direct conflict that many see between the Agile Manifesto and the need to have a fully documented and well-engineered system. Critical software systems cannot afford the high rate of defect generation that pure Agile executions output. Therefore, systems engineering must be incorporated into the development cycle of these systems.

The question then becomes, why do we not just use the normal, Predictive-style methodology of developing software systems if it is so critical? There is a logical reason behind that. The reason this is not done is that software projects have a greater ability, more so than any other engineering effort, to pivot and change requirements throughout their lifecycle. This is a superpower of sorts in the world of engineering management. There is no manufacturing plant, parts machining, logistics planning, or any other driving factor that makes up-front planning critical for traditional engineering activities. Due to this, maintaining flexibility in requirements and execution is paramount.

What then is the benefit of implementing systems engineering methods in Agile projects? Systems engineering brings full technical rigor to the software systems lifecycle, ensuring documentation and knowledge transfer throughout the life of the project. Systems engineering ensures a holistic approach, bringing a broader systems perspective with consideration of interaction between components, hardware, and human users. Thorough test and evaluation and requirements management are integral to systems engineering, ensuring that defects are minimized and that software is in a working and valid state prior to release.

In this chapter, we will discuss the benefits of taking a systems engineering approach to software development while ensuring that the flexibility inherent in the

D. Rosson, *Systems Engineering in Agile Software Development*,
https://doi.org/10.1007/978-3-032-14613-7_2

Agile process is not lost. Agile and Predictive development methodologies will be explained, along with the benefits and drawbacks of each. We will delve into requirements management, exploring why it is of absolute importance to critical software systems development. Finally, a case will be made for Agile User Stories to not be considered requirements at all in the traditional sense, but rather as Goals.

2.2 Overview of the Agile Scrum Development Methodology

Agile project management is a flexible form of project management, primarily used in software projects. The Agile Manifesto sets the baseline standards of what an Agile project should strive for. The original authors of the Manifesto strove to a better way of developing software. Using these principles, multiple ways of executing Agile projects have come into being since then, including Crystal, eXtreme Programming, Scrum, and numerous others. Agile practices have further evolved to program and business management through the utilization of scaling agile with methods such as Scrum of Scrums (SoS) or the Scaled Agile Framework (SAFe).

Agile project management is most useful when a project is high risk, low certainty, and requirements are only partially known. As shown in Fig. 2.1, agile scrum lowers project risk by employing an incremental delivery method and frequent communications and demonstrations with customers. This iterative and incremental approach helps to lower risk and optimize predictability through customer and subject matter expert engagement with the development team. By focusing on the most important need of a customer and then iteratively revisiting the project requirements, Agile methods make use of Lean thinking, reducing waste and focusing on what the customer needs, lowering overhead burdens.

While there are many forms of Agile program management, the most utilized is Scrum. Scrum is so widely used that many people assume that Scrum is the only form of Agile, so much so that practitioners have made many websites dedicated to explaining that Scrum is not the only Agile framework. The human mind struggles

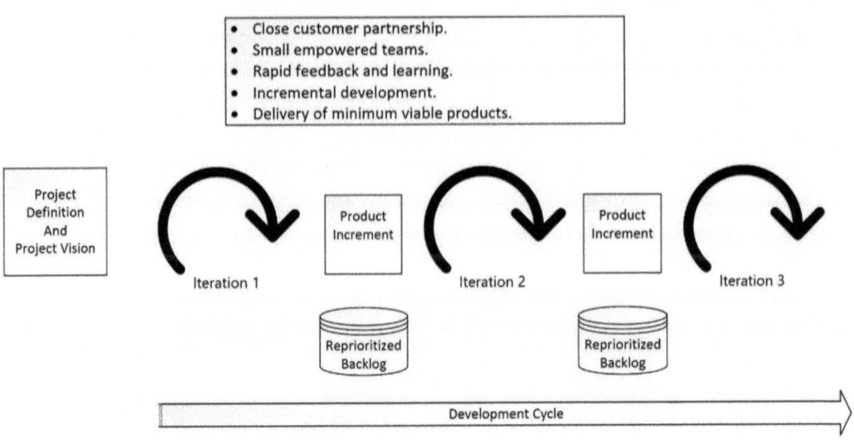

Fig. 2.1 Agile development approach

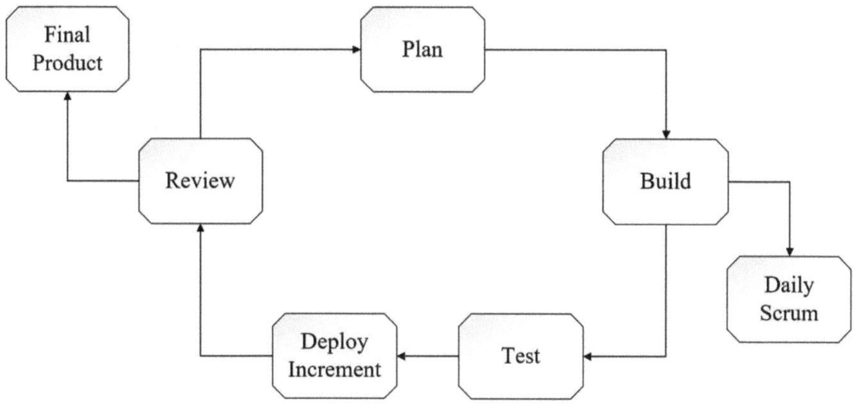

Fig. 2.2 The agile scrum execution cycle

to perform complex knowledge work in detail such as planning, understanding scope, task creation and assignment, consequences of actions across a system, and estimating complexity. Scrum assists with this effort by optimizing heuristics, providing continuous feedback, and creating external, cognitive artifacts, such as sticky notes, models, the Scrum Board, etc. Figure 2.2 shows an example of the Scrum product lifecycle. This lifecycle is a constant loop, progressing through the closed loop stages of Plan, Build, Test, Deploy Increment, and Review. During the Build Stage, a Daily Scrum activity occurs. After the Review stage, a Final Product might be produced if the Product Owner deems the product ready for production. This execution loop occurs each Sprint and continues until project completion.

Agile software development focuses on lightweight, flexible, and adaptable project execution. It is most useful when a project is considered high risk, low certainty, and requirements are only partially known. Agile Scrum further codifies the broader Agile methodology by implementing a structured framework which includes an iterative development cycle called a Sprint. Agile Scrum practitioners plan projects in Sprint increments, modifying execution plans as customer needs and requirements change. Through the utilization of Agile project management tools, Agile practitioners can swiftly adapt to changes to projects with a minor impact on cost, schedule, and performance.

2.3 Overview of the Predictive Development Methodology

Traditional, predictive project management is founded on a predictive lifecycle and has been around far longer than any other form of project management, having evolved during the 1950s. Predictive lifecycles are expected to have little flexibility or change over the course of execution, have readily available and stable team members, and accept negligible risks. These projects are most successful when the

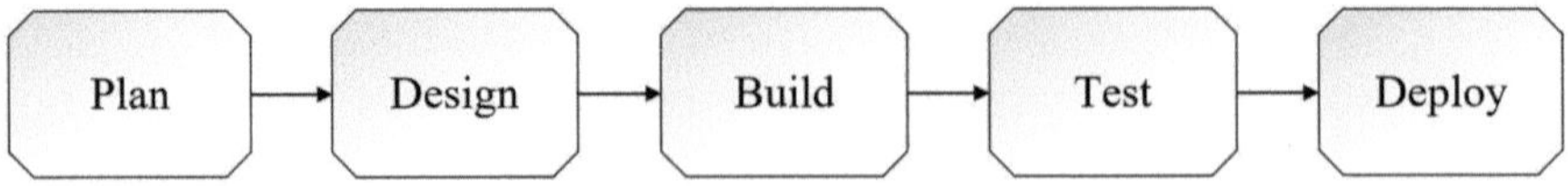

Fig. 2.3 Example predictive lifecycle

problem domain is clearly defined, and requirements can be generated upfront. For a project to be successfully executed using a predictive method, it must have a very well written set of requirements, a stable funding stream, a consistent and well-organized team, and high confidence that changes will not be forthcoming. This type of project management is well suited for construction, with a repeatable and sequenced progression of work. An example predictive lifecycle flow is shown in Fig. 2.3 where project phases are executed in a serial manner, starting with the Plan phase and working through all phases until the Deploy phase is completed. Many modern software projects are often have evolving project and product requirements, are high risk, and have significant uncertainty, making it hard to apply the predictive development approach to these types of projects successfully [1, 2].

The predictive methodology of project management is a linear and serial effort. The previous phase must be completed before moving to the next phase. Phases are not repeated, and work products should not be touched upon again. The only feedback loop that is present in this type of model normally occurs during the Test portion of the lifecycle. Defect fixes and changes can be approved and performed during this period. One of the major issues with this linear model is that when changes are requested or defects occur, there is a resultant extension of the project completion date, which creates Earned Value compromising problems.

Predictive and Agile methods are dramatically different and are best applied in different scenarios. Predictive methods are best suited to projects with minimal risk of change and uncertainty for which detailed requirements document can be generated. In these cases, detailed planning can lead to cost savings through the avoidance of unnecessary schedule slippage and cost planning. Predictive approaches are most suited to applications such as manufacturing and construction, where the final product is known, tolerances are documented, and the customer has a low likelihood of presenting changes to requirements. Agile methods are suited for medium to high-risk projects with a high degree of uncertainty where it is easy to pivot and course correct in the middle of project execution. In these types of projects, requirements do not need to be fully specified prior to the start of development. Instead, requirements can be elicited and documented at a higher level where customers' needs and goals are noted but not necessarily down to the atomic level. This is especially applicable in the software development industry, as producing code has little overhead cost outside of labor. Preplanning is not required for these projects, nor are there concerns of retooling, logistics and supply issues.

2.4 A Comparison of Requirements Management in Predictive Versus Agile Projects

Predictive-style requirements are described in INCOSE documentation [3]. INCOSE defines requirements utilizing specified language, such as a "shall" statement. These requirements are very detailed and must meet the following characteristics: necessary, appropriate, unambiguous, complete, singular, feasible, verifiable, validatable, correct, and conforming. Furthermore, the set of requirements must adhere to the following characteristics: complete, consistent, feasible, comprehensible, validatable, and correct. Requirements management in Predictive projects is, therefore, a detailed and large undertaking. The goal of an initial baseline requirements set is to have a fully formed vision and understanding of the proposed system.

In contrast, Agile DoD projects usually generate requirements in the form of Connextra User Stories. The Connextra form of User Story is written as the following: "As a role, I want goal, [so that][benefit]." These User Stories became popular through the ideas of eXtreme Programming, where development teams use User Stories to document customer needs from initial idea to the final product. These User Stories are stored by the developers in a backlog where they are prioritized by a Product Owner who sets priority according to their understanding of customer needs and desires.

User Stories are often tracked via a system called "Story Points." Story Points "rate the relative work, risk, and complexity of a requirement or story." As it is difficult to know exactly how much work, risk, or complexity a work item will have without breaking the item down into an atomic, easily testable, and understandable state, this can be a difficult task to accomplish. Many teams use a form of Story Pointing called Planning Poker to form a level set baseline for Story Point values. Each team has a different method to set point equivalence, and it can take time for the point values to stabilize with a new team. Experienced developers are of great assistance in this aspect, and it is recommended that teams be made up of junior and senior developers, as the collaboration of senior developers with junior developers increases the knowledge of both. Senior developers can pass on lessons learned from their long career while junior developers can teach the senior developers new techniques that they might not be familiar with. Furthermore, the complexity of estimation increases as groups can become over-confident in the accuracy of their estimates when given tasks are more difficult than are usually presented. Software developers often have problems predicting their own capability of work, and overall experience can be very "narrow." Therefore, while Story Pointing efforts can be effective in an experienced and fully realized team with mixed experience levels, there are valid concerns regarding the validity of storming to norming teams and their ability to estimate software project workloads adequately.

So, although both Agile and Predictive methods use these techniques to develop and communicate customer intent, in practice, they are not equivalently effective in various project contexts. It has been reported in numerous studies that Agile projects are reported as failures in by customers roughly 50% of the time within the bounds of system cost, schedule, and performance. Furthermore, for Agile projects to be successful, customers must actively participate in the process and are expected to work closely with the teams, represent their respective areas of responsibility, and be empowered to make authoritative decisions. Ineffective communication and poor requirements development are driving factors for failure in Agile projects.

The use of thorough, standardized, and atomic requirements in the form of predictive-style "shall" statements increase the ability of developers and customers to communicate with one another, directly leading to improved system and problem domain understanding. Especially in the instantiation phases of projects where Needs, Goals, and Requirements are generated—even if they are solely high-level and abstract. Because of the limited and abstract nature of User Stories, their use can lead to vague requirements as well as confusion within Agile project development teams. Problems with cost, schedule, and performance are even more likely to arise as availability of stakeholders is reduced.

2.5 A Comparison of Requirements Traceability

2.5.1 Introduction to Requirements Traceability

Requirement owners must trace requirements to needs for project success. Requirements traceability is defined as "the ability to describe and follow the life of a requirement in both forward and backward directions (i.e., from its origins, through its development and specification, to its subsequent deployment and use, and through periods of ongoing refinement and iteration in any of these phases)" [4]. Mature projects include traceability not only during one point in time, such as between customer requirements and system requirements, but also throughout time as requirements change and evolve. Through the application of traceability, the disparate artifacts that encompass a system are connected through linkages to other system components. Figure 2.4 shows how individual and disconnected system artifacts flow through a transformation activity where needs, goals, and requirements are traced directly to the artifacts, resulting in a network of system components.

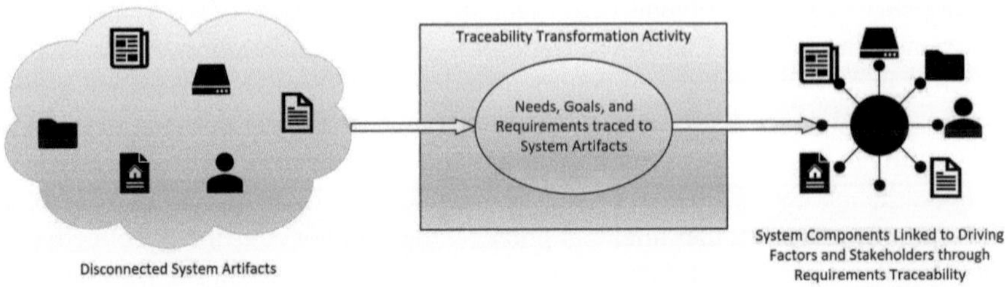

Fig. 2.4 The transformation of disconnected system artifacts to a requirements traced system

Maintaining traceability of system artifacts to individual requirements creates greater visibility and understanding of the system.

2.5.2 Agile Requirements Traceability

Many Agile Scrum teams recognize the importance of requirements traceability as a key enabler in maintaining customer focus during the software development process. Yet, requirements traceability is not inherent in Agile Scrum and is not mentioned in the Scrum Guide [1]. Agile projects generally track requirements at four levels, Epics, Features, User Stories, and Tasks. Many software developers do not hold traceability in high regard. It is often up to the individual team to determine if traceability is of value to their project or not. This represents the feeling of Agile Scrum practitioners where engineering and technical rigor are not normative, but instead only enacted if viewed as a net-gain in value. Due to this, traceability utilization beyond the hierarchy prescribed by basic User Stories is not standardized across the development industry.

The most basic hierarchy that is found in most Agile Scrum Product Backlogs follows the format of Epic → Feature → User Story → Task. Epics are the highest form of User Story representing an idea that is too large to comprehend by itself or more clearly, "a large User Story that cannot be delivered as defined within a single iteration or is large enough to be split into smaller User Stories." Many teams use Epics to track unrealized ideas within a Product Backlog prior to a team identifying the deliverable needed. Features are the next step down from Epics. The relationship between Epics and Features is a one-to-many relationship. The term Feature is not well defined but is generally realized as larger groupings of User Stories that are too big to be depicted as a single item, but too small to be considered Epics. User Stories clustered in this way are tied to specific common functionality. Below Features in the hierarchy is the User Story. The relationship between Features and User Stories is a one-to-many relationship. A described above, User Stories are an "end goal, not a feature, expressed from the software user's perspective" that is to be realized in the software system. Each User Story should be small enough to be completed within one Sprint cycle. In the most basic of terms, a User Story is something a user wants. Finally, User Stories can be further broken down into Tasks. The relationship between a User Story and a Task is a one-to-many relationship. This hierarchical traceability is shown in Fig. 2.5.

For example, agile scrum User Stories track traceability through natural language in the Connextra format. In the statement "As a {role} …," the role keyword

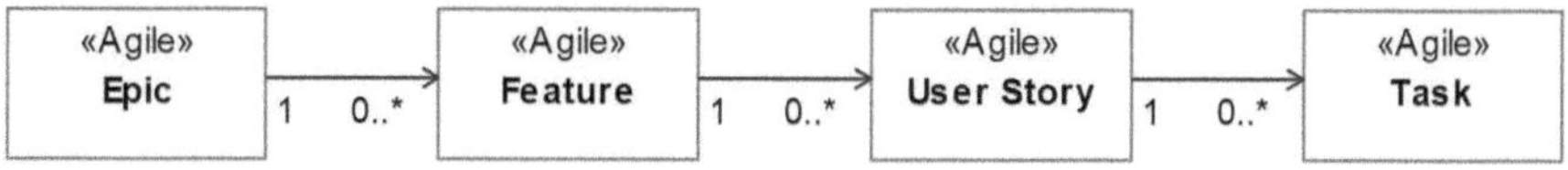

Fig. 2.5 Agile scrum user story hierarchy

is replaced by a user type. Many times, this is expressed as a defined "Persona." The purpose of the persona is to assist the Agile Scrum team in discovering the value of the product they are making. A Persona is a representation of a customer with specific needs or wants, such as a driver of a sports car versus the driver of utility truck. In other words, Personas are a fictionalized representation of a customer type that Agile Scrum software developers use to gain insight into customer needs, goals, and requirements. This traceability can be useful but is limited as it does not identify specific stakeholders as requirement owners. Rather, it identifies customer types, which software developers may or may not have access to for elaboration of needs, goals, and requirements. Agile Scrum teams that do not use Personas often do not have firm definitions of user roles, limiting traceability use.

Software developers have not settled on a single method for more rigorous and thorough requirements traceability. Some teams opt to use Excel style spreadsheet for requirements tracking, but this can lead to increased costs of traceability creation and maintenance. Other teams may opt for a project wiki to gather various artifacts, including requirements documentation, to allow for ease of editing as many wikis have versioning built into the core functionality. Various software applications have appeared on the market to solve the traceability issues inherent in software design, specifically with User Stories. Atlassian's Jira tool allows software developers to link work items such as User Stories and Tasks to test artifacts, defects, change requests, and other items. None of these methods are industry standard and teams use whatever is preferential internally.

Precluding organizational directives, each Agile Scrum team is free to cherry-pick what parts of User Story requirements traceability the team desires. Some Scrum Teams will ensure that there is as high a level of traceability as possible for their projects. Other teams will ignore traceability altogether, outside of the minimal traceability that User Stories have built in inherently. Therefore, there cannot be a definitive case set forth that the majority of Agile Scrum teams utilize good requirements traceability.

2.5.3 Predictive-Style Requirements Management Traceability

INCOSE emphasizes that traceability is "critical to managing relationships" and establishing traceability is an essential activity in the Technical Processes [3]. Establishing traceability is an essential activity in the Technical Processes as described in the INCOSE "Systems Engineering Handbook" [5]. In the INCOSE standard, traceability allows systems engineers to "understand then identify, location, relationships, pedigree, origin of data, materials, and parts of the objects/entities/items" [5]. It is therefore not surprising that the INCOSE methodology of requirements management has traceability inherent in the system. Requirements traceability takes many forms in the Predictive-style requirements management methodology.

Requirements, both as a set and individually, are traced both vertically and horizontally [3]. Vertical traceability is traceability of a hierarchical nature. Practitioners commonly refer to relationships of this type to as "parent–child" relationships. The parent is considered directly "above" the child in the hierarchy. Tracing requirements in this manner allows for a top-down perspective of a system and defines system hierarchical architecture. Horizontal traceability is used to trace requirement data through the system lifecycle, as well as across a given level of system architecture. Linkages of items generated in a lifecycle stage that are to be used by another lifecycle stage is an example of horizontal traceability. Horizontal traceability relationships are referred to as "peer" relationships. By tracing relationship connections both vertically and horizontally, a network representing the full view of the system of interest emerges, decreasing the effect of large and interconnected systems complexity.

Due to the way that requirements are documented in Predictive-style requirements management, any item can be traced to any other item. Documentation of linkages to other artifacts that each requirement traces to is part of a requirements register, linking each unique line item to other unique line items, as shown in Fig. 2.6. Thorough requirements traceability is not only encouraged but required in the technical processes. Without the built-in traceability, many requirements registers would be flat files and indecipherable to most people. Examples of common traceability found in Predictive-style requirements are parent-child relationships, source derivation, interfaces, and dependencies.

Model-Based Systems Engineering activities expand upon traceability in Predictive-style requirements management, directly increasing the level of visibility and understanding that requirements offer. As MBSE efforts, specifically SysML, rely heavily on Predictive-style requirements definition and standards, MBSE should be considered a part of Predictive-style requirements Management rather than a neutral third-party methodology. Requirements managed in SysML allow for additional traceability relationships realized through metadata connections. These relationships include, but are not limited to, derive, refine, satisfy, verify, and trace. The trace relationship is a generic relationship that is discouraged in favor of more precise relationships [6]. When making use of other MBSE tools such as model-based structured requirements (MBSR), even more detailed traceability is allowed

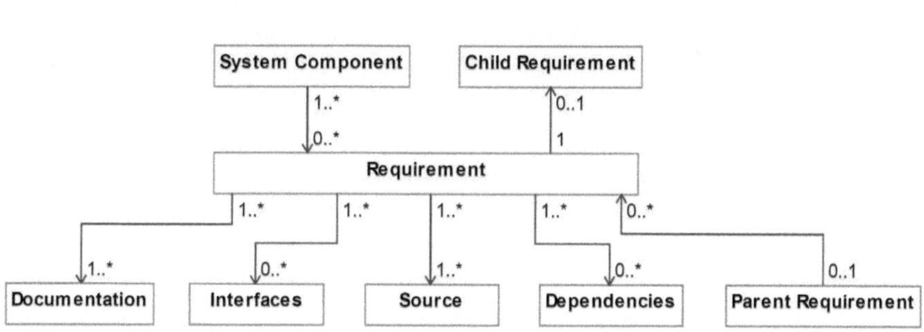

Fig. 2.6 Predictive-style requirements management traceability

for, such that attributes "can be linked to other model elements that represent aspects of the physical system, precise mathematical conditions, test cases, etc." through the use of stereotypes [7]. A stereotype is a form of metadata that defines how an existing meta class may be extended enabling the use of platform, domain specific terminology, or a notion in place of, or in addition to, the ones used for the extended meta class [8]. Furthermore, documenting the rationale of requirements is incredibly important to traceability and the understanding of the "why" of the requirement itself. A MBSE engineer can easily accomplish this through the <<rationale>> stereotype in MBSE linked to a textual comment detailing the rationale. With modeling, linkage relationships are easily understood by a larger audience as the media presented does not require mental modeling to visualize. Rather, the connections are plainly drawn in a graphical manner.

The INCOSE methodology of integrating traceability directly into requirements management in an open and flexible manner is much more robust and useful for system modeling and understanding complex systems. Predictive-style requirements traceability allows for modularity through the flexibility to allow a requirement to be traced to any system element, giving clear visibility into how requirements are being met. This visibility directly results in increased verification and validation capabilities through a clearer understanding of the system. Utilizing MBSE techniques, requirements can be traced directly to system architecture, better displaying how requirements are being realized through implementation. Tracing requirements back through derivations, sources, interfaces, documentation, and many other inputs allows systems engineers to fully conceptualize the problem space, building a mental model of not only what needs to be built, but also how and why each requirement is necessary.

2.5.4 Requirements Traceability Conclusions

Both Agile Scrum User Stories and Predictive-style requirements management have a form of traceability inherent in the system. The traceability is present in diverse ways in both systems. Predictive-style requirements management uses a structured, bidirectional, and hierarchical methodology that has firm rules that requirements engineers must follow [3, 5]. Agile Scrum User Stories allow for a more diverse inheritance process, but it is generally considered a good form to follow a top-down hierarchical structure. Requirements management style shall statements have inherent traceability to any other item that may affect them, whereas Agile Scrum User Stories are self-contained, except in cases where teams use metadata, which is not generally proscribed. Both styles of requirements have traceability, though predictive-style requirements management presents a more robust capability.

While both Agile Scrum User Stories and Predictive-style requirements management methodologies include built in traceability, User Stories' traceability is not as robust as requirements management style "shall" statements. User Stories have minimal traceability, prescribing only a "parent–child" relationship. This restriction can be overcome but requires outside intervention using such tools such as Jira. Predictive-style requirements management builds modular and flexible traceability into the system, allowing for traceability to any other requirement or system element. Not only does it allow for a greater range of traceability capability, but requirements management also dictates that all requirements should name types of traceability outside of the "parent–child" relationship, such as requirement derivation, verification methods, how requirements are refined, and how requirements are satisfied in the system. This allows for a more robust mental model to be built by the development team, facilitating an increased likelihood of delivering what the customer needs while minimizing rework that results in project cost, schedule, and performance slippages. It is clear that although User Stories do have minimal traceability inherent in the Connextra format, Predictive-style requirements management outperforms User Stories in every metric.

2.6 Goals Versus Requirements

2.6.1 Introduction to Goals

Goals are the "why" of the system. They are an expression of a desired outcome or intent and what the stakeholders want to achieve. Goals differ from requirements as they are abstract expressions of a customer desire for a system function or performance, where requirements are a much lower level, many times describing a functionality. Goals are normally vaguely formed and are tied to cost to benefit analysis and document stakeholder intention refining the system vision into objectives to be fulfilled. They are often qualitative, aspirational, and not directly verifiable.

Goals are commonly associated to scenarios in requirements management. Goals describe what the stakeholders want at a and scenarios describe situations in which the goal is used or needed. Goals and scenarios also work very well together to assist in the development of new requirements as stakeholders detail what the system "should" do. Goal traceability and relationships are shown in Fig. 2.7.

2.6.2 User Stories as Goals

User stories are not detailed enough to be considered true requirements, as they do not have all the metadata associated with a Predictive-style requirements management style requirement. The simple structure of the connextra user story format prevents a full understanding of the domain and system function and relations to be drawn from a user story. User stories are realistically goals and not requirements. In the connextra format, the word goal is used vice requirement, further reinforcing this. The ambiguity of User Stories, therefore, aligns with goals versus requirements.

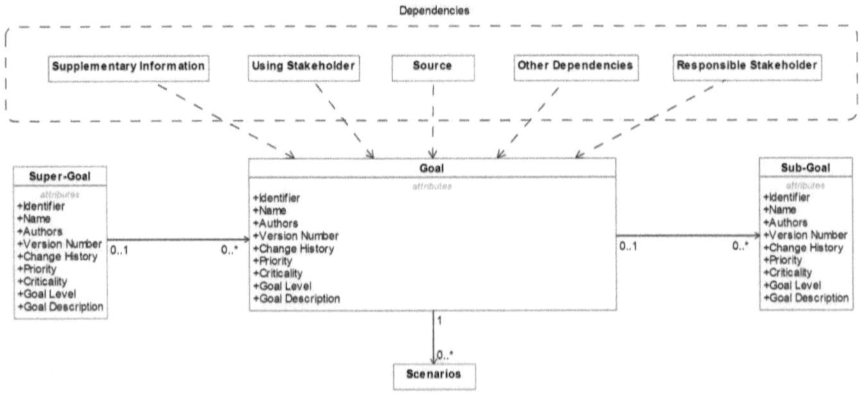

Fig. 2.7 Goals—traceability and relationships

As user stories are not detailed enough to be considered true requirements, documentation using solely ambiguous User Stories creates problems in understanding the proposed system domain and functionality. The problems, if left undetected, propagate throughout the Agile software development process and documentation, which leads directly to a result of negative affect on the system in development. By not detailing requirements at the lowest level possible, but instead relying on goals, software developers' risk not fully understanding the problem domain and therefore may not architect software correctly. This leads to waste in the form of rework that can dramatically affect cost, schedule, and performance in a project.

2.7 Systems Engineering in Software Conclusions

What are the interrelationships between project requirements and changes that drive project success when utilizing Agile Scrum User Stories versus systems engineering shall statements? There are many commonalities and differences between User Stories and Predictive-style requirements management style requirements. Predictive-style requirements offer more engineering and technical rigor through built in robust traceability and metadata. User Stories offer a high-level, abstract, and easy to execute system. Both forms of requirements management offer minimal traceability to user, or persona, types, needed function, and reasoning. User Stories are used in Agile software development and Predictive-style requirements are more often used in traditional, predictive projects. Agile Scrum User Stories should not be considered requirements when compared to the INCOSE definition. Instead, they are goals. They are abstract definitions of what a customer needs and wants in a system. Systems engineering shall statements are very concrete statements of what a customer needs a system to do. User Stories are too abstract to communicate customer needs while shall statements can be too rigid in management and tracking to allow for the flexibility software developers need to perform Agile development. The abstract nature of User Stories can directly cause the introduction of defects, both in requirements and in the system, through misunderstanding or miscommunication with the customer. In contrast to systems engineering shall statements, Agile User Stories do not have thorough traceability horizontally and vertically throughout the system life cycle. This creates difficulties in the performance of Verification and Validation activities, which leads to the introduction of defects. Both systems engineering shall statements and User Stories have benefits and drawbacks when applied to DoD Agile software projects and neither individually fully capture the needs of development teams with regard to executing projects.

Due to the lack of large-scale traceability and metadata linkages, User Stories alone do not meet the needs of critical systems engineering focused, integrated software projects. Predictive-style requirements management is not flexible on its own to allow ease of tracking in Agile Scrum Sprints, as requirements are considered too "low" in scope. On the basis of the above reflections on the current state of User Stories and Predictive-style requirements management, it is clear that a dual system is needed. Initial requirements elicitation through Predictive-style requirements management methods should be performed to create a requirements baseline. From this initial baseline, Epics can be drawn and traced directly to individual requirements. In this way, the flexibility of User Stories can be taken advantage of while the traceability and rigorous technical documentation of Predictive-style requirements management can still be employed, and the needs of critical systems can be met.

2.8 Discussion Questions

Perceived conflict: Where do the Agile Manifesto values appear to conflict with systems engineering rigor in critical software, and which misunderstandings drive that perception?

Why not purely predictive-style development: The text argues that software's unique ability to pivot makes pure predictive lifecycles ill-suited. Under what specific conditions would a predictive approach still be preferable, and why?

Rigor inside agile: What concrete systems engineering practices (requirements baselining, interface control, verification planning, hazard analysis, etc.) integrate cleanly into Scrum without negating agility?

User stories vs "shall" statements: Analyze User Stories as goals rather than requirements. What essential metadata and traceability do "shall" statements provide that User Stories typically lack, and how does that affect verification and validation?

Traceability depth: Compare Agile backlog hierarchy traceability (Epic, Feature, Story, and Task) with predictive bidirectional traceability (vertical and horizontal). How do these schemes change risk visibility over time?

Estimation and Story Points: What cognitive and organizational pitfalls affect Story Point estimation, and how can teams mitigate early "storming" inaccuracies while preserving useful forecasting?

Evidence of success and failure: The text cites mixed Agile outcomes and the need for active customer participation. What participation patterns, artifacts, and ceremonies most strongly correlate with successful delivery in high-risk domains?

MBSE's role: How do MBSE relationships such as satisfy, verify, refine, and derive enhance requirements comprehension and test planning compared to minimal trace links in many Scrum tools?

Dual system proposal: Critique the proposed dual system: predictive-style requirements baseline feeding Epics and a Scrum backlog. What governance checkpoints, tool integrations, and change-control rules are needed to keep both sides synchronized?

Cost, schedule, performance impacts: Using a critical system example, trace how ambiguous User Stories can propagate to cost growth, schedule slip, and degraded performance. Which specific systems engineering controls would break that failure chain?

References

1. Project Management Institute, *Agile Practice Guide* (Project Management Institute, Newtown Square, 2017)
2. Project Management Institute, *The Standard for Project Management and A Guide to the Project Management Body of Knowledge* (Project Management Institute, Newtown Square, 2021)
3. INCOSE Requirements Working Group, *Guide to Writing Requirements* (International Council on Systems Engineering (INCOSE), San Diego, 2023)
4. Chief Information Officer, DoDAF DoD Architectural Framework Version 2.02 (United States Department of Defense, 2010), https://dodcio.defense.gov/library/dod-architecture-framework/. Accessed 3 Mar 2024
5. International Council on Systems Engineering, *Systems Engineering Handbook*, 5th edn. (International Council on Systems Engineering (INCOSE), San Diego, 2023)
6. J. Wheaton, D.R. Herber, in Digital Requirements Engineering with an INCOSE-derived SysML meta-model. *Conference on Systems Engineering Research (CSER) 2024*, Tucson, AZ, 2024
7. D.R. Herber, K. Eftekhari-Shahroud, in Building a requirements digital thread from concept to testing using model-based structured requirements applied to thrust reverser actuation system development. *9th International Conference on Recent Advances in Aerospace Actuation Systems and Components*, Toulouse, France, 2023
8. No Magic, Inc., MagicDraw 19.0 LTR. Dassault systemes. https://docs.nomagic.com/display/MD190/Stereotype. Accessed 21 Apr 2024

Chapter 3
The Need for Engineering Rigor in Critical Projects

3.1 An Overview of Critical Software Systems (Fig. 3.1)

I have often referred to critical systems in this book, with only giving a high-level explanation of them. In Chap. 1, I defined critical systems as those in the realms of defense, medical, aviation, and finance. There obviously are many more fields that can be added under the umbrella of critical systems, such as IT, cybersecurity, and the automotive industry, to name a few. It should also be noted that not all systems in these program areas are critical systems. How do we further define a critical system to understand why a system is considered critical? To do this, we must expand the definition and drill down into what criticality is.

In the terms of a critical system, I posit that criticality is the level of effect that failures and defects in systems will have on stakeholders and customers if realized. Some systems, perhaps a small phone calendar application, can afford failures without drastically affecting the lives of the customer. If a defect is pushed to production, or perhaps the system has a catastrophic failure, lives will be affected, but not in a permanent manner. In other systems, such as an airplanes avionics system, a defect can be catastrophic, causing immediate danger to the lives of the operators and people onboard. Therefore, we will define critical systems as systems in which a defect or failure has the potential for a catastrophic effect on the customer or stakeholder with regards to life, health, or finances.

Now that we have defined a critical system, the definition of a critical software system becomes much clearer. A critical software system is a software system, embedded, web, traditional client/server, or mobile app, in which a defect or failure has the potential for a catastrophic effect on the customer or stakeholder with regards to life, health, or finances. It is important to remember that software is rarely deployed as an individual system. Even web-applications are hosted in a server environment, even if they are a virtual cloud based one, and must be treated and

D. Rosson, *Systems Engineering in Agile Software Development*,
https://doi.org/10.1007/978-3-032-14613-7_3

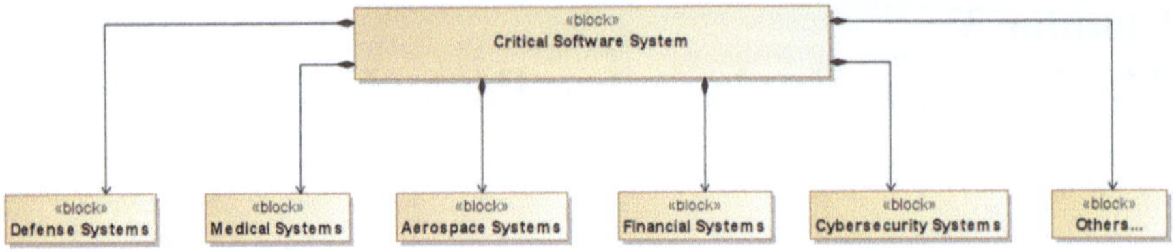

Fig. 3.1 Critical software systems

thought of in a systems engineering manner, as whenever multiple systems interact, emergent behaviors occur.

With these definitions in mind, it is easy to see that many software systems can be considered critical. There are even systems out there that industry does not realize would fall within the bounds of this definition as they are not the full system itself, but rather a part of a System of Systems. In these cases, defects and poor performance can have an adverse effect on the System of Systems, created negative emergent behaviors. The problem is complex and hard to grasp, but it is one that must be taken into consideration.

3.2 Challenges in the Application of Agile in Critical Systems Development

Having defined what constitutes a critical software system, it is now pertinent to transition from theoretical considerations to practical application. Conceptual understanding offers a valuable foundation; however, meaningful advancement is achieved by analyzing the tangible challenges encountered throughout the design, development, and maintenance phases. These issues are expressed through the complex interplay of performance requirements, safety obligations, cost limitations, and project timelines.

As discussed in Chap. 2, applying Agile principles without modification may result in challenges when working with critical systems. The use of Agile emphasizes speed and flexibility, which can introduce risks related to cost, schedule, and performance, not to mention long-term system stability. However, we should not reject the use of Agile but rather acknowledge that all methodologies should be adapted to suit their respective domains. In high-risk contexts, such as critical system development, alignment between method and the systems mission is essential to avoid undesirable outcomes.

In the following pages, we will explore the challenges of the application of Agile development methods in critical system development in greater depth. We will examine how the drive for rapid iteration, to "go fast and break things," can conflict with the need for engineering and technical rigor, how software systems' ever-changing requirements can undermine architectural stability, and how cultural divides between traditional systems engineers and Agile practitioners can actually slow progress, not speed it up. Each of these issues represents not merely a technical hurdle, but a systemic challenge that demands thoughtful application of engineering rigor and Agile practices. By studying these challenges closely, we will better understand how to transform these potential risks into structured opportunities for innovation and resilience, creating better software systems that with far more stability and customer satisfaction.

3.2.1 *Safety and Mission Assurance*

The top and highest priority for any critical system is ensuring safety first and mission assurance second. When developing a system utilizing an Agile methodology, the emphasis is put on the rapid output of working software. In critical systems, safety, security, and mission assurance must be guaranteed prior to deployment. This is yet another area in which the motto of "go fast and break things" does not apply. A single defect, if pushed through to a production environment, could disable a Navy ship's navigation system, cause accounting errors in a bank's financial transaction tracking system, or cause fatal crashes in aircraft. In these types of environments, the stakes are not just inconvenient but rather existential where failures mean catastrophic consequences for customers, stakeholders, operators, and sometimes entire corporations and nations.

Defects in critical systems are not just inconveniences for the system users, but can result in system failure, loss of life, or mission compromise. With this in mind, additional layers of rigor are necessary above and beyond what is normal in the Agile execution environment. Increment development must focus as much on fault containment as it does on features. Thorough requirements management must be performed, including accounting for Non-Functional Requirements, such as safety, security, and reliability. These requirements are often pushed to later Sprints but must not be deferred. Security is as paramount as safety. Formal verification and validation must be performed. Test-driven development alone utilizing automated unit tests may not capture all failure conditions and edge cases. Formal test methods, including tracking test coverage, test to requirements traceability, simulation, and exhaustive verification must accompany automated testing.

The promise of Agile's rapid delivery and adaptability is true and valuable, but in critical software systems development, it must be bound within the confines of considerations for safety and mission assurance. The speed of development and delivery of increments must not justify the cutting of corners with regard to reliability, security, and fault tolerance, because even small defects can potentially lead to catastrophic outcomes if released into production. Engineering and technical rigor are not optional, but rather are the foundations that validate every increment against life-critical and mission-critical requirements. Agile activities must then be adapted to embed assurance activities into each cycle, such as simulation, formal testing, and thorough requirements management. Software development must operate in service of quality and safety, rather than undermining and working against it, while still ensuring responsiveness to potentially changing customer needs.

3.2.2 Traceability and Requirements Management

As was shown in Chap. 2, Agile methods emphasize the management of evolving requirements through User Stories, customer collaboration, and backlog refinement. While this flexibility is powerful in commercial software development, especially in the startup industry, it clashes with the traceability demands of critical systems designed with good engineering and technical rigor. In defense, medical, aerospace, and other critical domains, requirements must be clear and traceable from stakeholder intent all the way to system implementation and test results. If this chain is broken, defects can be introduced and functionality not fully realized, resulting in customer and stakeholder loss of confidence at best and catastrophic failures at worst.

In Agile, User Stories and Epics often represent requirements and needs, respectively. We have already addressed how this is problematic in Chap. 2, as User Stories are actually more closely aligned with the INCOSE definition of a goal. Agile practitioners would argue against that and state that User Stories are intentionally left flexible to enable change. In critical systems, this leads to problems. First, to be able to fully test a requirement, it must be clear, concise, represent one concept, and be atomic. Oftentimes, User Stories require additional metadata to meet this standard, which leads to testability issues. Second, bidirectional traceability is imperative in requirements management for critical systems. User Stories have no inherent traceability outside of the serial Epic → Feature → User Story → Task. There are no linkages to test cases, implementation, or rationale. This is left up to the developer to guess at. Finally, there must be change control mechanisms to ensure modifications are thoroughly tested and confirmed before release to a production environment. While it is true that not all Agile development teams practice rapid deployment methods, such as Continuous Integration/Continuous Deployment (CI/CD) or DEVSECOPS, many do, and it is important to note that and to perform controlled release planning. The development flows, or pipelines, in these types of methods focus on small increments and continuous delivery to customers, as can be seen in Fig. 3.2. In cases where CI/CD is used by the development team, perhaps releases to an integration server for full testing and baselining prior to release to production is warranted to minimize impacts of defect release and understand the traceability of changes to requirements, lowering the risk of unintentional changes being introduced into production. As can be seen, Agile teams that rely only on evolving backlogs of User Stories risk requirements drift, introduction of technical debt, loss of accountability, and the release of unintended code into production environments, all of which are unacceptable in critical systems.

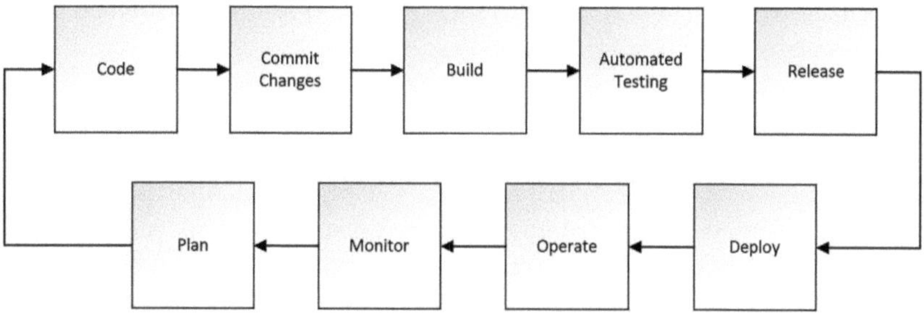

Fig. 3.2 CI/CD pipeline

3.2.3 Governance, Compliance, and Standards

The development of critical systems often falls under the umbrella of strict regulations and certification frameworks. Some examples of these frameworks are the DoD 5000 acquisition series, the Federal Aviation Authority (FAA) DO-178C for airborne systems, and the NASA Procedural Requirements (NPRs). These frameworks have been put in place to reduce program and system risk where failure could lead to catastrophic results for the mission, major financial setbacks, or, at worst, the loss of life. The frameworks enforce rigor in planning, verification, and documentation to ensure that every system and subsystem correctly follows and aligns with mission and safety objectives. These regulations focus on traceability and formal compliance artifacts through strong use of engineering and technical rigor, which stands almost opposed to the values of the Agile manifesto. This creates tension between traditional engineering teams and Agile focused software developers. One methodology embraces and thrives on change while the other desires stability and consistency, desiring control throughout development. If not addressed, Agile practitioners will feel slowed and unsatisfied by the regulatory demands put upon them or be out of step with the expectations of oversight bodies.

It can be difficult to cleanly map Agile ceremonies and artifacts into the formal milestones, compliance checkpoints, and contractual deliverables required by regulations. While a Sprint Review often demonstrates working software, certification boards and milestone review panels often require tangible objective quality evidence (OQE) in the form of detailed requirements matrices, test plans and results, and design documentation. In fact, with the release of the DoD 5000.97 "Digital Engineering" instruction, all defense programs are to implement digital engineering practices, including thorough Model-Based Systems Engineering and the creation and maintenance of digital threads. This perceived misalignment can create disconnects between what the Agile team views as forward momentum and what the oversight authorities recognize as necessary rigor for system stability. Moreover, Agile artifacts are typically generated for communication, both intra-team and external to the team, not for audit or formal decision authority. As a result, project managers often find themselves duplicating efforts through the creation of transitional artifacts, one set for the Agile team and another for compliance reporting, which can increase their workload and slow the overall delivery. Overtime, this forced compliance can undermine Agile's promise of efficiency if not carefully managed.

Many of these types of programs utilize Earned Value Management (EVM) and milestone reviews to assess project health focusing on a cost versus schedule or product delivered versus Agile's focus on delivered value. This can increase the communication difficulty of Agile project managers as most Agile projects do not have a set deliverable path that aligns with the principles of EVM. Moreover, planning milestone reviews at a specified date places an artificial "done" date on the Agile team, restricting them from delivering early if needed. Agile teams tend to focus on tracking project execution using velocity charts, burndown graphs, and backlog trend analysis, all of which focus on value delivered to the customer and not "earned value," or value traced to the dollar spent. Attempting to force Agile health assessments into EVM can create situations where system development looks healthy in Agile metrics but is underperforming in EVM. For example, EVM would call for a constant execution rate and a stable velocity. This is often not the norm in Agile teams, especially Agile teams that have come together for a new project. Large amounts of learning and team "norming" happens in the beginning of the project, which leads to lower velocity numbers. The middle of the project might see high velocity numbers, as the team has become efficient working together and are outputting the easy and average difficulty functions while researching more difficult functionality. Toward the end of a project, velocity might drop as harder functionality is finally implemented. It could even increase at this time as overtime is authorized, and the team performs swarming[1] activities to ensure a final release date is met. Velocity is not directly tied to hours worked; therefore, it would create an artificial dip or rise in the EVM metrics. As is clear, it is a difficult problem to align EVM and Agile execution. The burden of reconciling this disparate view often falls to project and program leadership, who must then translate Agile artifacts into compliance ready metrics. Without a plan and template to transition these metrics, the project risks being seen as non-compliant or out of scope with relation to cost, schedule, and performance, despite making real progress.

Finally, compliance auditors and milestone authorities often rely upon technical documentation to prove regulatory compliance, traceability, and verification and validation. As previously noted, the Agile Manifesto specifically calls out for a focus on delivered and working software over documentation, which most Agile teams use as an excuse to minimize documentation. Agile views unnecessary documentation as a hindrance to development. However, in critical systems, documentation is not optional. It serves as the historical record of these systems for future auditors to track compliance against and maintainers and certifiers depend upon. If managed in a fully Agile manner, critical systems may fail certification gates even though they passed in earlier looks. This can lead to program delays or, in the case of defense acquisition programs, cancelation. Then challenge then lies in creating

[1] In software development, swarming is an agile technique where multiple team members, with the necessary skills, focus on a single high-priority task or user story until it is fully completed.

documentation that satisfies the needs of oversight and compliance while also not overburdening the Agile teams. Successful programs achieve this through embedding documentation tasks in the Agile workflow and the utilization of MBSE, ensuring that necessary compliance evidence is produced and delivered incrementally along with the software.

How does all of this look in practice? Critical systems managed in Agile environments must embed compliance, audit, and assurance artifact generation into the workflow so that velocity can be maintained while also meeting regulatory requirements. Programs that are successful in this treat compliance not as a burden to project execution, but rather as a form of quality assurance built-in to the workflow. This requires close collaboration between the Agile team, the quality assurance team, project management, and the compliance auditors to collaboratively create and agree upon compliance ready artifacts to meet milestone requirements. Digital engineering tools support this by automating traceability between user stories, requirements, test cases, and architecture. Through this alignment of Agile with regulatory compliance, a more stable system is developed and the spirit of Agile can be preserved, ensuring a fast-moving and trustworthy environment and satisfaction from stakeholders.

3.2.4 Cultural and Organizational Resistance

The culture of traditional engineers is grounded in the principles of thoroughness before action. Those with a traditional engineering background are trained from the beginning to define, analyze, and document every requirement and design element prior to committing to implementation. This working belief has its roots in the concept that a problem discovered late in development is exponentially more expensive to fix than one that was identified and solved earlier in the lifecycle. As a result, programs that are managed in a more traditional engineering project management style emphasize exhaustive upfront analysis, formal systems engineering technical reviews, and meticulous documentation. This approach is performed to enhance stability and program manager confidence, especially in domains where safety, mission assurance, or public trust are at stake, such as in critical systems. Aerospace, defense applications, and nuclear systems owe much of their reliability to this disciplined approach. Yet, this structure can also be a limitation. Comprehensive planning often slows down agility and adaptation. The pursuit of perfect foresight limits timely learning.

Early software execution was attempted under this same Predictive style. It was often not successful and stymied the creativity and adaptability that is inherent in software development. Due to this, the various Agile methods emerged as a counterbalance to Predictive rigidity. Where traditional engineering methods seek to control the process, Agile focuses on flexibility, discovery, and learning. It assumes that change is inevitable, that requirements are evolutionary, and that problem domain knowledge grows not through prescription, but through natural experimentation.

These teams learned to focus on functional value, learning from each iteration and adjusting along the way. They prioritized customer interaction, collaboration, and incremental progress. This adaptability proved to be transformative. Utilizing Agile, teams can respond to new information without derailing an entire project or rewriting large swaths of documentation. However, when applied to critical systems with high-assurance requirements, this flexibility can result in unsafe deliveries and unacceptably low understanding of the problem domain. Without some form of deliberate discipline, the same speed that fuels the Agile innovation will also amplify the inherent risks.

The difficulty then is the reconciliation of these two distinct development views and is one of the central challenges in developing critical software systems. This is both a philosophical divide as well as a procedural one. Traditional engineering teams trust compliance and planning whereas Agile practitioners trust the speed and feedback loops. In critical systems where failure is not an option, neither side can completely dominate the other and a compromise must occur. Rigid plans result in the loss of the ability of software to adapt to changing and growing customer needs, leading to development teams delivering products that don't fully meet the needs of the customer. On the other hand, a focus on iteration and Agility without engineering and technical rigor accounted for leads to instability in the product, often times with a severe price paid in program cost, schedule, and performance. Creating a merger of these two philosophies requires the implementation of structure with adaptability and the creation and acceptance of a culture in which documentation supports learning rather than impeding it and iteration serves quality rather than undermining it. This is not something that management can just write into policy, but rather it is a cultural change that must occur. The merger must focus on a shared language through requirements, integrated processes, and finally, trust between the disciplines. When this is performed successfully, the result is a development culture that moves with speed and Agility, but never at the expense of safety and innovation.

3.2.5 Risk Management and Fault Tolerance

In critical system development, risk management and fault tolerance are the cornerstones of engineering rigor. Whereas Agile frameworks emphasize adaptability and rapid iteration, critical systems demand exhaustive analysis to identify, classify, and mitigate hazards before the first line of code or hardware prototype is built. In such environments, even a single overlooked failure mode can have catastrophic consequences leading to unacceptable risks to safety, financial stability, and mission success. Therefore, early integration of systematic risk identification methods such as Preliminary Hazard Analysis (PHA) and Failure Mode and Effects Analysis (FMEA) is essential to ensure that adaptability does not come at the expense of safety or mission assurance.

A well-structured risk management plan anchors the system design process in evidence-based decision making. Each identified risk must be categorized by its likelihood, severity, and detectability, establishing clear mitigation paths and verification criteria. Fault tolerant design, in turn, transforms risk mitigation into architectural practice by embedding redundancy, graceful degradation, and recovery mechanisms directly into the system's architecture. This ensures that even if one component fails, the overall system continues to function safely and predictably. In contrast, Agile teams operating without these frameworks often defer such analysis to later iterations, inadvertently compressing the time available for fault detection and remediation.

Agile's focus on delivering visible, user-facing functionality can lead to an underemphasis on latent system level reliability. Iteration-driven sprints naturally prioritize demonstrable progress, often measured through features and user stories, rather than through less tangible metrics like mean time between failures (MTBF) or fault coverage ratios. As a result, crucial reliability engineering practices such as fault-tree analysis (FTA), reliability block diagrams (RBD), and stochastic modeling can be sidelined in favor of velocity. While this approach may accelerate short-term development, it undermines long-term system resilience and increases the likelihood of costly post-deployment failures.

To reconcile these competing priorities, Agile teams developing critical systems must embed risk management activities within their iterative cycles. This includes integrating hazard analysis and failure mode modeling as recurring sprint deliverables through integrated unit testing or manual test execution, not as one-time gate reviews. Automated reliability simulations, digital twin environments, and model-based safety analyses can provide near real-time feedback on system robustness as features evolve. By coupling Agile's responsiveness with rigorous model-based verification, teams can maintain both adaptability and reliability without deferring safety considerations until it is too late in the lifecycle.

Architectural foresight remains the decisive factor in achieving fault tolerance, one of which Agile teams do not necessarily consider until it is "just in time." Early architectural design decisions have an outsized impact on system resilience. Examples of this are choosing between centralized and distributed control, defining redundancy levels, or determining communication failover protocols. Neglecting these decisions during early iterations can create single points of failure that no amount of later testing can fully resolve. Consequently, architects of critical systems must treat fault tolerance planning as an integral design objective rather than a secondary optimization. The cost of redundancy is often negligible compared to the cost of catastrophic failure.

Moreover, organizations developing critical systems must institutionalize continuous risk governance. Risk registers, safety case documentation, and formal hazard logs should evolve alongside the system, ensuring that every iteration accounts for emerging risks and design changes. This institutional rigor transforms risk management from a compliance activity into a dynamic engineering discipline. Metrics such as fault exposure rate, recovery time objective (RTO), and system availability provide quantitative feedback loops that reinforce decision-making across development and sustainment phases.

Ultimately, risk management and fault tolerance form the bridge between Agile adaptability and engineering discipline. Critical systems cannot rely solely on emergent design or reactive mitigation; they demand proactive architecture, continuous validation, and verifiable redundancy. By embedding structured risk analysis into iterative development, engineers can preserve the agility needed to innovate while upholding the safety and reliability standards that protect missions, assets, and lives. The balance between adaptability and rigor defines not only technical success but also ethical responsibility in the engineering of critical systems.

3.2.6 Schedule and Cost Pressures

For Agile teams working on critical systems, the tension between Agile adaptability and fixed contractual baselines is particularly pronounced. Agile methodologies thrive on flexible scope, iterative planning, and continuous reprioritization; yet in domains such as defense, aerospace, and medical systems, cost and schedule are often rigidly defined through acquisition contracts, milestone gates, and government oversight. These projects operate under strict accountability frameworks where deviations from approved baselines must be formally justified and reauthorized. The result is an inherent cultural and procedural friction. There is a general view that Agile's responsiveness clashes with the immovable constraints of contractual engineering.

In these environments, cost and schedule baselines are legally binding commitments tied to funding authorities, program offices, and regulatory compliance. A single iteration that extends beyond the contracted budget or milestone window can trigger cascading impacts: withheld payments, formal corrective action requests, or mission-critical delays. Such outcomes are not simply administrative burdens; they can jeopardize deployment readiness, disrupt downstream integration activities, and erode sponsor confidence. For systems whose delivery is synchronized with fleet schedules, launch windows, or clinical trials, even minor deviations can have outsized operational consequences.

Agile teams may find these constraints stifling as they are accustomed to re-estimation and scope adjustment as tools of transparency and control. In commercial software, adjusting velocity or backlog scope midstream is a mark of responsiveness. In defense systems acquisition, it can be perceived as a lack of discipline in planning. When program managers or contracting officers encounter repeated re-baselining, they may interpret it as cost growth or lack of control rather than adaptive management. This perception is particularly damaging in environments where auditability, traceability, and earned value metrics form the foundation of program oversight. Therefore, it is understandable that contracted agencies are reticent to embrace Agile methods if doing so will potentially harm their ability to do business.

To navigate this divide, programs overseeing critical systems must hybridize Agile execution with traditional systems engineering governance. This often takes the form of tiered control gates where high-level cost and schedule baselines remain fixed, but lower-level Agile iterations operate flexibly within defined "control accounts." Under this model, teams can adjust sprint-level priorities while maintaining cumulative adherence to contractual cost and milestone targets. Such governance structures preserve both transparency and predictability, translating Agile adaptability into a language that aligns with acquisition oversight and earned value management (EVM) frameworks.

In practice, this hybrid approach is most effectively supported through cost-plus contracting mechanisms, which provide the flexibility necessary for iterative development while maintaining oversight of total program expenditure. Unlike fixed-price contracts which can penalize adaptive re-estimation or evolving technical requirements, cost-plus contracts allow the contractor to recover allowable costs and earn a negotiated fee, accommodating controlled scope evolution as system understanding matures. This structure fosters collaboration between the contracting agency and the development team, incentivizing transparency, continuous communication, and shared accountability for technical outcomes. In critical system programs, cost-plus contracting thus becomes a key enabler of disciplined agility through preserving the benefits of iterative refinement without compromising the fiscal control demanded by government or regulatory bodies.

Risk-based planning and continuous forecasting become essential tools in managing this balance. By integrating cost and schedule performance indicators thoughtfully into Agile program reviews, teams can identify deviations early and implement corrective actions before thresholds are breached. Traditional performance metrics such as the Cost Performance Index (CPI), Schedule Performance Index (SPI), and burndown projections can be successfully adapted to provide quantitative oversight without sacrificing iterative flexibility. The key lies in translating Earned Value Management (EVM) concepts, which were originally designed for linear, Predictive style projects, into metrics that operate on a sprint or increment basis rather than a full project phase and tied directly into a quantifiable metric, such as story points completed versus planned. When properly implemented, these indices enable program managers to track progress and efficiency while maintaining alignment with contractual cost and schedule baselines. Model-Based Systems Engineering (MBSE) further enhances this alignment by linking requirements, schedules, and costs within a single digital framework, ensuring that changes to one element propagate transparently across the entire system model. This digital thread strengthens both adaptability and compliance.

However, even with these integrations, cultural perception remains a persistent challenge. Traditional stakeholders may continue to view Agile as "undisciplined" due to its iterative nature and evolving scope, especially in the realm of defense and aerospace industries. Overcoming this bias requires deliberate communication strategies that demonstrate control through evidence: traceable metrics, real-time dashboards, and documented risk mitigation actions. When program leaders can visualize progress, earned value, and technical performance within each iteration, the narrative shifts from perceived chaos to demonstrable control.

In the end, schedule and cost pressures are not obstacles to Agile, they are the constraints within which true engineering discipline is proven. Critical system development must treat cost and schedule adherence as performance parameters no less vital than safety or reliability. Engineers can show that flexibility and rigor are not opposites but complements by embedding financial and temporal accountability into the Agile cadence. The goal should not be to abandon adaptability, but rather to institutionalize it within a disciplined, transparent, and contract-compliant framework that satisfies both innovation and assurance in the most demanding environments.

3.2.7 System of Systems Complexity

It is rare in modern engineering that a system exists as a standalone entity. Systems often operate within expansive system-of-systems (SoS) that encompass interconnected platforms, data networks, sensors, and control systems. In defense, aerospace, and medical domains, success depends not only on the performance of each subsystem but on the seamless interoperability of all components functioning as a coherent whole. Aircraft communicate with satellites, hospitals exchange patient data across regional networks, and undersea platforms rely on secure coordination with surface and space assets. This interdependence introduces layers of emergent behaviors, or outcomes that cannot be predicted or validated solely through the lens of local subsystem performance.

While highly effective for driving rapid iteration, Agile's team-centric model often narrows the focus to local Sprint goals and immediate deliverables. Teams measure success through completion of user stories or feature increments, sometimes without full awareness of the system's broader architectural dependencies. In a system-of-systems context, such local optimization can yield global degradation. A feature that improves efficiency in one subsystem might introduce timing conflicts, data latency, or communication bottlenecks when integrated into the enterprise ecosystem. This "local agility versus global stability" tension lies at the heart of SoS engineering challenges.

Emergent behaviors, defined as behaviors that arise only when systems interact, represent one of the most significant risks in engineering complex architectures. These behaviors often cannot be detected through isolated testing or modeled through linear cause-and-effect assumptions. For instance, two individually stable control loops may oscillate when coupled, or security measures in one domain may create unintended vulnerabilities in another. Without mechanisms to simulate and analyze these interactions early, programs risk discovering incompatibilities only during late-stage integration, when remediation is most costly and disruptive.

A lack of coordination across Agile teams amplifies this risk. When multiple teams develop components concurrently with minimal synchronization, interface definitions, communication protocols, and timing constraints can drift. Integration then becomes an exercise in reactive troubleshooting rather than disciplined synthesis. The absence of shared architectural baselines and system level verification models allows misalignments to compound over time. The result is often an accumulation of "boundary risk." These are defects or inconsistencies that emerge precisely where systems meet. In critical systems, boundary failures can cascade, leading to mission degradation or total system outage, as has happened ever more frequently in the past decade, leading to numerous major incidents.

To mitigate these risks, SoS engineering within Agile environments requires architectural governance and cross-team synchronization mechanisms. Practices such as System Integration Reviews (SIRs), Model-Based Interface Control Documents (MB-ICDs), and digital thread architectures provide a unified framework for validating interoperability at each iteration. By maintaining a shared model that links local development activities to enterprise-level behaviors, programs can continuously assess emergent interactions and trace the impact of local design decisions across the broader system. This enables teams to adapt locally while staying aligned globally.

Furthermore, system architects and chief engineers must serve as integration stewards, ensuring architectural coherence while facilitating Agile flexibility. These leaders ensure that each team's backlog items are not only technically feasible but contextually compatible within the SoS environment. Integration testbeds, simulation environments, and digital twins can provide virtual proving grounds where emergent behaviors are observed before physical integration occurs. Through these practices, fault detection and system validation become continuous processes rather than end-of-phase events.

Ultimately, managing system-of-systems complexity in Agile development demands both structural and cultural discipline. Teams must view their work not as isolated increments but as interdependent threads in a larger operational tapestry. Engineering rigor serves as the connective tissue that binds local agility to enterprise reliability. Without it, even the most innovative Agile implementations risk unraveling under the weight of their own interdependencies.

3.2.8 Documentation Debt and Knowledge Capture

Among the most persistent challenges in applying Agile to critical systems is the issue of documentation debt. Documentation debt is the cumulative shortfall that arises when teams postpone detailed documentation in favor of near-term delivery, as is extremely common in Agile teams. Agile philosophies often promote the notion of "just enough documentation," emphasizing working software and direct communication over comprehensive recordkeeping. While this approach accelerates early development and reduces administrative overhead, it poses serious risks in environments where sustainment, certification, and compliance depend on complete, traceable, and enduring documentation. In critical domains, the absence of formal records is not merely an inconvenience but rather represents a fundamental breach of engineering rigor.

Defense, aerospace, and medical programs operate under regulatory frameworks that require verifiable documentation of every phase of the system lifecycle. Design rationale, requirements traceability, configuration baselines, and verification evidence are all things that must be auditable, reproducible, and retained for decades. These artifacts enable certification authorities to confirm that systems meet safety and mission assurance standards. When Agile teams defer or minimize documentation, the resulting gaps can delay certification, invalidate compliance audits, or force costly retroactive reconstruction of design decisions. In high-stakes systems, documentation is as much a safety mechanism as a recordkeeping practice.

Moreover, documentation debt compounds invisibly over time. A single sprint's skipped design note, in-line code documentation, or omitted test rationale may seem trivial, but across multiple iterations and teams, the cumulative effect erodes institutional memory. Without continuous documentation updates, models lose synchronization with code, system interfaces drift from specification, and the authoritative source of truth fractures. Reconstructing these links later consumes exponentially more effort than maintaining them incrementally. Thus, the true cost of documentation debt is not in the time saved upfront, but in the time saved performing rework and in the uncertainty and risk introduced downstream.

Another critical factor in amplifying this risk is personnel turnover. Defense and aerospace programs often span decades, far exceeding the tenure of individual engineers or contractors. When institutional knowledge resides primarily in the minds of team members rather than in structured artifacts, its loss through attrition becomes inevitable. The departure of a single subject matter expert can leave critical gaps in understanding, particularly for systems with proprietary interfaces, complex integration histories, or legacy technologies. Without well-maintained design records and knowledge repositories, sustainment teams are left to reverse-engineer decisions made years and sometimes even decades earlier.

To address these challenges, Agile programs developing critical systems must institutionalize knowledge capture as an engineering function, not an afterthought. Model-Based Systems Engineering (MBSE) provides a powerful foundation for this effort by embedding documentation directly into the system model itself. In a model-centric environment, requirements, interfaces, behaviors, and verification data are interlinked and automatically updated as designs evolve. This eliminates the disconnect between evolving code and static documentation, ensuring that the model remains the living representation of the system's current state.

In addition to MBSE, organizations should adopt structured knowledge management frameworks including decision logs, design rationale repositories, and digital engineering playbooks to codify reasoning behind key architectural choices. These mechanisms allow future engineers to understand not only what decisions were made but why. Embedding knowledge capture into the definition of done for each sprint ensures that documentation evolves concurrently with functionality, transforming it from a compliance burden into a natural artifact of progress.

Ultimately, documentation in critical systems is not antithetical to agility. It is the foundation that sustains it over time. Agile practices must evolve to recognize that "just enough" documentation for a short-lived commercial product is inadequate for systems expected to operate safely for decades. In high-consequence domains, documentation is both a safeguard and a legacy. It preserves institutional memory, supports certification, and enables future innovation. The disciplined capture of knowledge, when integrated into the Agile cadence, ensures that adaptability today does not come at the cost of comprehension tomorrow.

3.2.9 Conclusions

The application of Agile in critical systems development presents a paradox: the same qualities that make Agile successful in commercial software such as speed, adaptability, and iterative delivery can, if left unchecked, undermine the rigor, stability, and assurance demanded in safety- and mission-critical domains. Throughout this section, we have explored how the principles of Agile must be adapted, not abandoned, to meet the stringent expectations of defense, aerospace, medical, and other high-reliability environments. Challenges such as ensuring safety and mission assurance, maintaining end-to-end requirements traceability, meeting governance and compliance standards, managing cost and schedule pressures, and mitigating risks through fault tolerance all demonstrate the importance of integrating flexibility within a disciplined engineering framework.

Agile's strengths lie in its responsiveness and its focus on delivering incremental value, but these qualities must operate under the governance of structured engineering processes. Critical systems cannot afford uncontrolled experimentation or incomplete verification; every iteration must advance both functionality and assurance. As demonstrated, concepts like Model-Based Systems Engineering (MBSE), Earned Value Management (EVM), and cost-plus contracting provide the mechanisms to harmonize iterative progress with the rigor of regulated development. Similarly, practices such as continuous risk assessment, architectural foresight, and disciplined knowledge capture transform Agile from a purely adaptive framework into a controlled, evidence-based methodology suitable for complex, high-stakes environments.

Perhaps the most significant insight is that success in critical Agile programs depends not solely on process, but on culture. It requires engineers, managers, and stakeholders to move beyond traditional dichotomies and instead pursue an integrated philosophy of disciplined agility. This mindset accepts that rigor and adaptability are not competing forces but complementary ones. A system developed under this philosophy evolves responsively while remaining grounded in safety, traceability, and mission assurance.

Ultimately, applying Agile to critical systems is an act of engineering maturity. It demands that we retain the spirit of Agile innovation while embedding it within the proven safeguards of systems engineering. When executed with discipline, transparency, and foresight, Agile methods can deliver the adaptability needed to meet evolving mission requirements without sacrificing the stability upon which lives, assets, and national security depend. In this synthesis of rigor and responsiveness lies the future of critical systems development: fast when possible, deliberate when necessary, and uncompromising where safety and assurance are at stake.

3.3 The Systems Engineering Focused Agile Development (SEFAD) Process

3.3.1 Introduction

How can Agile Scrum project implement Systems Engineering methodologies without losing the ability to take direct customer feedback and requirements changes to meet a customer driven definition of done? To answer this question, a new methodology is needed which maintains the ability to quickly pivot as customer feedback is given and assessed, considering frequent changes in requirements, while still tracking requirements and documenting system state. The proposed process should focus on technical rigor at the beginning of the project while initial capabilities are generated and requirements gathered, transition to a flexible and highly Agile methodology during software execution, allowing for swift changes in requirements and architecture, and then transitioning back to a rigorous framework to close out the project. This will allow software developers to communicate with cross-functional teams and customers effectively while still working in an environment that can be adapted easily to changing customer driven requirements.

Through the data collected in reviewing critical software systems development case studies and through the knowledge gained from the Agile Scrum simulation models, it is apparent that the most influential systems engineering methods are the following:

1. Requirements management: Performing structured, systems engineering style, requirements management increased understanding of the system context as well as facilitating customer communication.
2. Customer collaboration: Creating, instilling, and ensuring a close and collaborative relationship between stakeholders and development engineers directly increases understanding of the system context and problem domain, leading to a more successful product viewed through the lens of customer satisfaction.
3. Documentation and technical rigor: Ensuring documentation and tracking of needs, goals, and requirements, as well as other technical documents, increases understanding of the system context and problem domain, directly increasing the ability for development engineers to execute new projects and sustain existing projects.
4. Modeling: Implementation of Model-Based Systems Engineering activities, such as SySML and Wireframe Modeling, increases the ability of stakeholders and developers to communicate ideas and collaborate on functionality, increasing the understanding of systems context and the problem domain.

Data collected from software projects executing in an environment that utilizes systems engineering methodologies along with the Agile Scrum Framework has shown that the development team gains a higher understanding of customer needs, goals, and requirements. This directly leads to an increase in system context and problem domain knowledge, as well as facilitating clearer communication with stakeholders. The goal in software projects is to deliver a product with limited scope creep and within budget while lowering stress on the development team and customers. The thoughtful merger of systems engineering methods with the Agile Scrum Framework is an enabling factor for this.

3.3.2 Systems Engineering Focused Agile Development (SEFAD) Process

The following process was developed to create a project execution environment merging the rigor of systems engineering methodologies with the flexibility of the Agile Scrum Framework.

3.3.2.1 Process Objective

1. To establish a process applicable to engineering activities across various programs or projects.
2. To analyze customer needs and translate needs into requirements.
3. To design, develop, and implement solutions to meet the customer's needs and requirements. Solutions will encompass products, product components, and product-related lifecycle processes either singly or in combination as appropriate.
4. To assemble and integrate the solution into a deliverable, supportable, and well-engineered product.
5. To verify and validate product requirements through rigorous testing

3.3.2.2 Process Boundaries

1. This process begins with the project initiating phase to ensure process flow and continuity throughout the product lifecycle.
2. The process concludes with product delivery and project closeout.

3.3.2.3 Background

The Systems Engineering Focused Agile Scrum (SEFAD) process is a software development methodology that leverages the rigor of systems engineering methodologies, such as comprehensive requirements management, with the flexibility and adaptability of Agile Scrum. Systems Engineering is a transdisciplinary and integrative approach to enable the successful realization, use, and retirement of engineered systems, using systems principles and concepts, and scientific, technological, and management methods. Agile Scrum is an empirical process, where decisions are based on observation, experience, and experimentation. Scrum has three pillars: transparency, inspection, and adaptation. This supports the concept of working iteratively. This process includes all stages of the product lifecycle, from initiating phase until sustainment phase. Although it is outside the scope of this directive, it is recommended that programs have a Systems Engineering Plan (SEP).

To better communicate the goal of SEFAD, the development of a system follows a rigid process from instantiation through requirements and architecture development, as shown in Fig. 3.3. Once the PDR has been successfully accomplished, the process transitions into a flexible, Agile development cycle, which continues until the system has passed the CDR. Once the CDR is completed, all requirements are baselined, and the system process once again, it becomes rigid. No new requirements will be added. Remaining unaccomplished requirements are developed and defects fixed. This continues until the system is ready for the Production Readiness Review and passing into delivery.

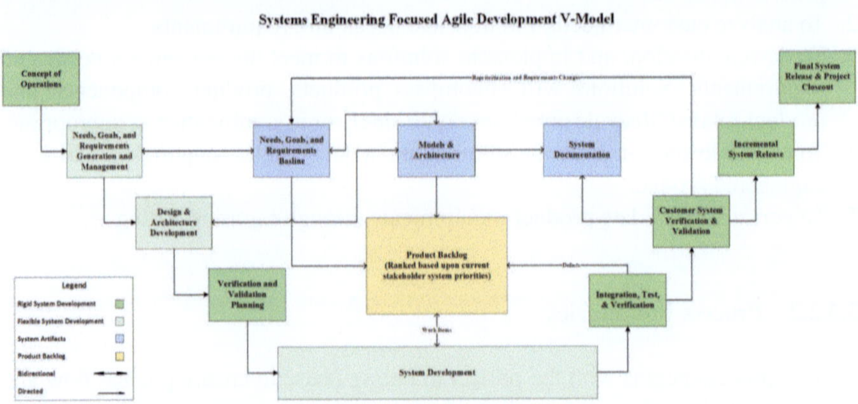

Fig. 3.3 SEFAD V-model illustrating rigidity and flexibility in software development

3.3.2.4 Policy

1. All major technical reviews will be conducted according to guidance from the prime stakeholder or customer. For example, if the project is under the purview of the DoD PEO IWS 5 program office, then the use of the PEO IWS Technical Review Manual must be adhered to. Stakeholders may use technical review guidance documents more specific to the design/development work being accomplished.
2. Tailoring Guidelines: This framework dictates the minimum effort required for development of systems. Some projects, being more complex than others, may add items, but all items as detailed in this document shall be completed for projects. The entrance and exit criteria for the decision points shall be documented in every case, tailored to fit the project, and shall become a project artifact that is signed off by an "Approving Authority" and retained as a permanent record for the project in accordance with the Configuration Management (CM) Plan.

3.3.2.5 Definition and Key Terms (Table 3.1)

3.3.2.6 Responsibilities

Stakeholder/Program Office

- Develop initial needs and high-level requirements.
- Provide guidance throughout the development process to ensure the derived requirements satisfy the high-level requirements and customer needs.
- Monitor project to ensure completed on time and within budget.

Project Manager (PM) or Delegated Authority

- PM leads decision point reviews and signs decision memorandums if the customer has delegated authority.
- Schedule and host decision point reviews.

Systems Engineer (SE)

The Systems Engineer develops, designs, allocates, and manages stakeholder needs and system level requirements, leads the development of the system architecture, evaluates design tradeoffs, balances technical risk between systems, defines and assesses interfaces, provides oversight of verification and validation activities, as well as many other tasks throughout the course of a program.

Table 3.1 SEFAD definition and key terms

CCB	Configuration control board
CDR	Critical design review—decision point (system level)
CM	Configuration management
Decision point	Administrative/technical review of project performance based upon fixed entrance and exit criteria. Culminates with a decision memorandum to continue to the next phase or iteratively rework the design
DDP	Design-dependent parameter. Design characteristics inherent in the product or system such as reliability, maintainability, and disposability
DIP	Design independent parameter. Design characteristics that are not controlled by the product or system such as labor rates, material costs, energy costs, etc.
FDR	Final design review
GUI	Graphical user interface
IDR	Initial design review
MVP	Minimum viable product. The minimum product that can be delivered to a stakeholder to meet stakeholder requirements
Need statement	A stakeholder's need is a high-level representation of a capability or objective the stakeholder would like to accomplish. It establishes the relationship between the organization and the customer. It is something that the stakeholders want or which the stakeholders feel is necessary. It is written in plain English and utilizes the keyword "need" Example: "The business needs to be able to generate reports that detail our revenue stream so that we can report to our board of directors concerning the company's profitability"
PDR	Preliminary design review—decision point (system level)
PO	Product owner
PRR	Production readiness review—decision point
Requirements statement	Requirements statements represent things that are needed to be accomplished to achieve a stakeholder need or goal. They are written in "shall" statements and can be broken down into lower-level increments to give further clarification. A requirement should state who shall (do, perform, provide, weight, or other verb) followed by a description of what should be performed and why it is being performed, if known Example: "The software shall generate reports that detail the company's revenue stream so that the customer can meet board of director mandated reporting requirements"
SE	System engineering. A transdisciplinary and integrative approach to enable the successful realization, use, and retirement of engineered systems, using systems principles and concepts, and scientific, technological, and management methods
SEFAD	Systems engineering focused agile development
SNR	Stakeholder needs review
SND	Stakeholder needs document. An official statement set of stakeholders needs that have been communicated by the stakeholders to represent what stakeholders want, or feel is necessary, to accomplish their goals and meet their objectives
SM	Scrum master
SME	Subject matter expert

(continued)

Table 3.1 (continued)

SRD	System requirement document. An official statement set of system requirements that have been negotiated and agreed to for stakeholders, customers, end-users, and clients
SRR	System requirements review—decision point (system level)
PM	Project manager
PM-A	Technical performance measure—attribute. A technique that measures the risks inherent in a technical system element to determine how well that element is satisfying requirements. This normally is the method of measurements of performance at specific points in a program and the variation limits for corrective action
WBS	Work breakdown structure
Work item	Decomposed elements of a project plan that are tasked to project members for completion

The Systems Engineer's responsibilities include:

- Developing and executing the system CDR plans with established quantifiable review criteria, carefully tailored to satisfy program objectives.
- Ensuring the pre-established review criteria have been met to ensure the design has been captured in the allocated baseline and initial product baseline.
- Ensuring assessments and risks associated with all design constraints and considerations are conducted, documented, and provided (e.g., HSI, reliability and maintainability, corrosion, SS, and ESOH considerations).
- Updating RIO plans. Identifying, analyzing, mitigating, and monitoring risks and issues and identifying, analyzing, managing, and monitoring opportunities.
- Monitoring and controlling the execution of the CDR closure plans.
- Documenting the plan to SVR in the SEP and elsewhere as appropriate.

Product Owner

The Product Owner is accountable for maximizing the value of the product resulting from the work of the Scrum Team.

Conduct overall day-to-day management of the project to include the following:

- Develop, maintain, and explicitly communicate the Product Goal.
- Create and clearly communicate Product Backlog items.
- Order and prioritize Product Backlog items.
- Ensure the Product Backlog is transparent, visible, and understood.
- The product owner is one person, not a committee. The product owner may represent the needs of many stakeholders in the product backlog. Those wanting to change the product backlog can do so by trying to convince the product owner.
- Translate customer/program office strategies and needs to workable tasks for development.
- Maintain knowledge of the market and customer needs.

- Serve as a liaison between the Customer/Program Office and the development team.
- Be accessible to the development team to answer questions related to the product.

Scrum Master

The Scrum Master is accountable for establishing Scrum as defined in the Scrum Guide. They do this by helping everyone understand Scrum theory and practice, both within the Scrum Team and the organization. The Scrum Master is also accountable for the Scrum Team's effectiveness. They do this by enabling the Scrum Team to improve its practices, within the Scrum framework.

The Scrum Master acts as a servant leader by serving the team's needs rather than commanding them, fostering an environment for them to be successful. Key aspects of this leadership style include facilitating events, removing impediments, coaching, and mentoring the team, and protecting the team from distractions. The focus is on the team's growth and effectiveness, not the Scrum Master's personal gain or authority.

The scrum master serves the scrum team in several ways, including:

- Coaching the team members in self-management and cross-functionality.
- Helping the Scrum Team focus on creating high-value Increments that meet the Definition of Done.
- Causing the removal of impediments to the Scrum Team's progress.
- Ensuring that all Scrum events take place and are positive, productive, and kept within the timebox.

The scrum master serves the product owner in several ways, including:

- Helping find techniques for effective Product Goal definition and Product Backlog management.
- Helping the Scrum Team understand the need for clear and concise Product Backlog items.
- Helping establish empirical product planning for a complex environment.
- Facilitating stakeholder collaboration as requested or needed.

The scrum master serves the organization in several ways, including:

- Leading, training, and coaching the organization in its Scrum adoption.
- Planning and advising Scrum implementations within the organization.
- Helping employees and stakeholders understand and enact an empirical approach for complex work.
- Removing barriers between stakeholders and Scrum Teams.

Developers

Developers are the people in the Scrum Team that are committed to creating any aspect of a usable Increment each Sprint. The specific skills needed by the Developers are often broad and will vary with the domain of work.

The Developers are always accountable for:

1. Creating a plan for the Sprint.
2. Creating and maintaining the Sprint Backlog.
3. Instilling quality by adhering to a Definition of Done.
4. Adapting their plan each day toward the Sprint Goal.
5. Holding each other accountable as professionals.

3.3.2.7 Reviews

Stakeholder Needs Review (SNR)

The Stakeholder Needs Review is a multi-disciplined technical review of the stakeholders' needs. The goal of this review is to ensure that all initial stakeholder needs are documented and that the project is ready to move on to system requirements definition.

Question to answer: Are all stakeholders' needs accounted for in the Stakeholder Needs Registry and is the system ready to develop system level requirements?

System Requirements Review (SRR)

The SRR is a multi-disciplined technical review to ensure that the developer understands the system requirements and is ready to proceed with the initial system design.

Question to Answer: Are all stakeholder needs accounted for in the System Requirements Document and is the system ready to develop system level requirements?

NOTE for U.S. Defense Acquisition Programs: A SRR is mandatory per DoDI 5000.88, Section 3.5.a. This review assesses whether the system requirements as captured in the system performance specification (sometimes referred to as the System Requirements Document (SRD)):

- Are consistent with the preferred materiel solution (including its support concept).
- Are consistent with technology maturation plans.
- Adequately consider the maturity of interdependent systems.
- Are clearly stated and are measurable and testable.

Initial Design Review (IDR)

The IDR is the first lightweight review held with the stakeholders. At the IDR, the Stakeholder and Product Owner will review the SND and SRD, as well as initial system models and T&E plans. After a review of the initial system design, the decision will be made as to whether the system as proposed will meet stakeholder needs.

Question to answer: Will the proposed system as currently designed meet the stakeholders needs?

NOTE for U.S. Defense Acquisition Programs: This should be held in the form of a formal PDR, mandatory for MDAPs per DoDI 5000.88, Section 3.5.a. The PDR should provide sufficient confidence to proceed with detailed design. The PDR determines whether the preliminary design and basic system architecture are complete, that there is technical confidence the capability need can be satisfied within cost and schedule goals, and that risks have been identified and mitigation plans established. It also provides the acquisition community, end user, and other stakeholders with an opportunity to understand the trade studies conducted during the preliminary design and thus confirm that design decisions are consistent with the user's performance and schedule needs and applicable requirements documents. The PDR also establishes the allocated baseline.

- See the Systems Engineering Guidebook for all required PDR activities [1].

Sprint Review

The purpose of the Sprint Review is to inspect the outcome of the Sprint and determine future adaptations. The Development Team presents their work to key stakeholders and progress toward the Product Goal is discussed.

During the event, the Scrum Team and stakeholders review what was accomplished in the Sprint and what has changed in their environment. Based on this information, attendees collaborate on what to do next. The Product Backlog may also be adjusted to meet new opportunities, as well as the Stakeholder Needs Register and System Requirements Document. The Sprint Review is a working session, and the Scrum Team should avoid limiting it to a presentation.

Publish Authorization Review

The purpose of the Publish Authorization Review is to inspect the current outcome of the Sprint and the state of the System Increment. The Development Team presents their work to key stakeholders with the goal of a decision to publish the current System Increment.

A Sprint Review is part of the Publish Authorization Review. The Development Team and stakeholders review what was accomplished in the Sprint and what has changed in the system environment. Based on this information, attendees collaborate on what to do next. The Product Backlog may also be adjusted to meet new opportunities as well as the Stakeholder Needs Register and System Requirements Document.

With stakeholder agreement and approval, a release publication authorization form is signed authorizing the release of the current System Increment. The approved System Increment is published to the system production environment. This is especially useful if utilizing a CI/CD pipeline and has a high chance of returning much needed stakeholder feedback.

Final Design Review (FDR)

At the FDR, the Stakeholder and Product Owner decide whether the system has reached a critical point of completion in that the pathway to final product delivery is obvious. Depending on the project, this could happen as early as 75% completion to as late as 95% completion.

Question to answer: Is the system sufficiently complete and documented to move into Project Finalization?

NOTE: This will result in a Stakeholder Needs and System Requirements freeze. No new work items, other than defects or emergent changes, shall be added to the Product Backlog for work

NOTE for U.S. Defense Acquisition Programs: This should be held in the form of a formal CDR as is mandatory for MDAPs per DoDI 5000.88, Section 3.5.a. The CDR confirms the system design is stable and is expected to meet system performance requirements, confirms the system is on track to achieve affordability and should-cost goals as evidenced by the detailed design documentation and establishes the initial product baseline.

- See the Systems Engineering Guidebook for all required CDR activities [1].

Production Readiness Review (PRR)

At the PRR, the Stakeholder and Product Owner decide whether the system is ready for final product delivery.

Question to answer: Is the system sufficiently complete that stakeholder needs are met satisfactorily enough to finalize the project and move to sustainment?

NOTE: This does not mean that all requirements have been developed or that the product backlog is empty. This means that the stakeholder is satisfied with the current state of the system as a working viable product. Remaining requirements and work items will be triaged and packaged to be transferred for sustainment efforts

NOTE for U.S. Defense Acquisition Programs: The PRR, mandatory per DoDI 5000.88, Section 3.5.a. determines whether the system design is ready for production, and whether the developer has accomplished adequate production planning for entering Low-Rate Initial Production (LRIP) and Full-Rate Production (FRP).

Production readiness increases over time with incremental assessments at various points in the program life cycle.

- See the systems engineering guidebook for all required PRR activities [1].

3.3.2.8 Process Flow

The process flow is highlighted in the following four figures. Note that these are typical engineering review Decision Points and Work Items. The intent is not to attempt to address every conceivable combination of tailorable Decision Points and Work Items but to provide a basis for informed decisions. The bottom line is the desire for a comprehensive, iterative, recursive, and well documented system engineering problem solving process. Immediately following the SEFAD process diagrams given is Table 3.2 which contains detailed information for each of the items addressed in the process diagrams.

3.3.2.9 Process Flow Details

Table 3.2 provides details on the process flow outlined in Figs. 3.4, 3.5, 3.6, 3.7, and 3.8. Figure 3.3 provides definition for the process flow charts.

- Individual Responsible: provides the Individual Responsible for accomplishing the task. In most case this will require bring together a Team of Subject Matter Experts (SMEs).
  - NOTE: Stakeholders may assign a singular individual or a group of individuals to represent their needs in this process. This authority shall be documented in the project plan.
- SMEs Involved: provides a list of SMEs that may be brought together to accomplish the task. This is just a typical list of the individuals responsible for particular actions. The Project Lead may obtain the services of anybody required to accomplish the task.
- Step: provides a pointer to the process flow figure step.
- Action: provides a high-level summary of the action required.
- Artifact: provides a list of the typical outputs associated with the action performed (Table 3.3).

3.3.3 SEFAD Implementation

The SEFAD process has been implemented successfully in multiple DoD and government projects. To implement a process that utilizes the application of systems engineering principles, project teams require training to understand the importance

Table 3.2 SEFAD process flow details

Individual(s) responsible	SME(s) involved	Step	Action	ARTIFACT
Product owner Stakeholder	SE PO TPM	10	Establish project records. Determine configuration management policy for project	• Initial project records • Configuration management policy
Systems engineer Product owner	SE PO T&E Development team	20	Develop stakeholder needs. A need is a high-level representation of a capability or objective that the stakeholder would like to accomplish. Needs are written as statements. Example: Customers need mobile devices that are efficient to install, operate, and maintain, so that they can spend their time doing things which are more important to them.	• Stakeholder needs registry
Systems engineer Product owner Stakeholder	SE PO T&E Development team	30	Validate the customer needs as documented in the stakeholder needs registry.	• Validated stakeholder needs registry
Product owner Stakeholder	SE PO	40	Are all stakeholder needs accounted for? Yes: Move to stakeholder needs review. No: Return to developing stakeholder needs.	• Stakeholder needs completeness decision
Product owner Stakeholder	SE PO TPM	50	Decision point: SNR Question to answer: Are all stakeholder needs accounted for in the stakeholder needs registry and is the system ready to develop system level requirements?	• Approval decision for stakeholder needs registry
Product owner Stakeholder	SE PO TPM	60	Yes: Transition to developing system requirements. No: Return to developing stakeholder needs.	• SNR pass decision

(continued)

Table 3.2 (continued)

Individual(s) responsible	SME(s) involved	Step	Action	ARTIFACT
Systems engineer Product owner	SE PO T&E Development team	70	Develop system requirements. A requirement is: 1. A condition or capability needed by a user to solve a problem or achieve an objective 2. A condition or capability that must be met or possessed by a system or system component to satisfy a contract, standard, specification, or other formally imposed documents 3. A documented representation of a condition or capability as in (1) or (2). Requirements are written as shall statements. Example: The system shall require users to enter a pin number prior to the access of system functionality	• System requirements document
Systems engineer Product owner	SE PO T&E Development team	80	Identify technical performance measures, design-dependent parameters, and design independent parameters.	• Updated system requirements document with TPMs, DDPs, and DIPs
Systems engineer Product owner Stakeholder	SE PO T&E Development team	90	Validate the system requirements as documented in the system requirements document	• Validated systems requirements document
Product owner Stakeholder	SE PO	100	Are all system requirements accounted for? Yes/pass: Move to system requirements review No/fail: Return to developing system requirements	• System requirements completeness decision
Product owner Stakeholder	SE PO TPM	110	Decision point: SRR Are all system requirements accounted for in the system requirements document and is the system ready to move into modeling and test planning?	• Approval decision for the system requirements document and initial baseline
Product owner Stakeholder	SE PO TPM	120	Yes/pass: Transition to the modeling and test planning phase No/fail: Return to developing system requirements	• SRR pass decision

(continued)

Table 3.2 (continued)

Individual(s) responsible	SME(s) involved	Step	Action	ARTIFACT
Product owner Systems engineer	SE PO Development team T&E	130	Perform initial modeling. • Wire diagrams • SysML • DoDAF	• Initial system models
Product owner Stakeholder	SE PO TPM	140	Verify the initial design: Question to answer: Does the proposed system, as shown in through initial modeling, meets customer needs and system requirements	• Verification of system models and initial design documents
Product owner Stakeholder	SE PO TPM	150	Yes: Proceed to the next steps No: Return to perform initial Modeling, taking into account stakeholder feedback	• Initial model approval decision
Product owner Systems engineer Test engineer	SE T&E	160	Create an initial test plan: • Create test plan • Create test matrix • Create initial test cases, if applicable	• System test plan • System test matrix • Test cases
Product owner Technical project manager	PO TPM	170	Create the development schedule • Create software development plan	• Finalized project plan • Software development schedule
Product owner Stakeholder	SE PO TPM	180	Decision point: IDR Question to answer: Will the proposed system meet stakeholder needs?	• Approved project plan • Approved initial test plan and associated artifacts • Approved initial development schedule • Approved initial system models
Product owner Stakeholder	SE PO TPM	190	Yes/pass: Transition to the development phase No/fail: Perform design verification	• IDR pass decision
Product owner Scrum master Systems engineer	PO SM SE Development team Logistics	200	Agile planning • Finalize sprint schedule, including sprint time boxes • Assign scrum master • Standup scrum issue tracking system • Create procurement plan, if necessary	• Agile Scrum schedule • Agile team roster • Software/ hardware procurement plan

(continued)

Table 3.2 (continued)

Individual(s) responsible	SME(s) involved	Step	Action	ARTIFACT
Product owner	PO SM SE Development team T&E	210	Create and manage the product backlog • Break down requirements into workable issues • Assign point values to the work items • Prioritize the product backlog • Perform Sprint planning	• Initial product backlog • Populated sprints
Product owner	PO SM SE Development team T&E	220	Perform sprints • Software development • Modeling • Database development • Procurements • Testing	• Demonstration of current product development through a Sprint review
Product owner	PO SE T&E Development team	230	Does the team feel an increment of the system is ready for release and delivery to the stakeholder? Yes: Hold a publish authorization review No: Hold a Sprint review	• Publishable increment decision
Product owner Stakeholder	PO SE T&E Development team	240	Sprint review • Demonstration of Sprint work accomplished • Determine if the system is ready for FDR	• Sprint closeout presentation • Record of accepted Sprint items • System ready for FDR determination
Product owner Stakeholder	PO SE T&E Development team	250	Publish authorization review • Demonstration of Sprint work accomplished • Determination if the system increment is ready for release • Stakeholders sign off on publish authorization form, if publish is authorized	• Sprint closeout presentation • Record of accepted Sprint items • Publish system increment decision. • Publish authorization form, potentially • Published system increment, potentially • System ready for FDR determination

(continued)

Table 3.2 (continued)

Individual(s) responsible	SME(s) involved	Step	Action	ARTIFACT
Product owner Stakeholder	PO SE	260	Permission to publish the system increment? Yes: Proceed to publish the system increment No: Proceed to system ready for FDR determination	• System increment publish decision
Product owner	PO SE T&E Development team	270	Publish the approved system increment to the production environment • Inform stakeholders of release publication date and potential downtime • Ensure training is updated • Snap a baseline for the published system	• Published system increment • Published system baseline
Product owner Stakeholder	SE PO	280	Is the system ready for FDR? Yes: Move to final design review No: Proceed to requirements review	• System ready for FDR decision
Product owner Stakeholder	SE Development team T&E	290	Perform a requirements review • Are the current requirements sufficient to represent the stakeholder's needs? • Does the stakeholder have new requirements? • Are there any requirements that are OBE?	• Requirements to be added • Requirements to be modified • Requirements to be removed • Needs to be added • Needs to be modified • Needs to be removed
Product owner	SE PO	300	Are there changes to requirements or stakeholder needs? Yes: Update the system requirements documents and/or the stakeholder needs register No: Proceed to create and manage the product backlog	▪ Requirements change decision.
Product owner	SE PO	310	Update system requirements and/or stakeholder needs • Update the system requirements to add, remove, or modify requirements as applicable • Update the stakeholder needs to add, remove, or modify needs as applicable	• Updated system requirements document • Updated stakeholder needs register

(continued)

Table 3.2 (continued)

Individual(s) responsible	SME(s) involved	Step	Action	ARTIFACT
Product owner Stakeholder	SE PO TPM	320	Decision point: FDR Is the system sufficiently complete and documented to move into project finalization? NOTE: This will result in a stakeholder needs and system requirements freeze. No new work items, other than defects or emergent changes, shall be added to the product backlog for work	• Approved project plan • Approved test plan and associated artifacts • Approved development schedule • Approved system models
Product owner Stakeholder	SE PO TPM	330	FDR passed? Yes/pass: Transition to the project finalization phase No/fail: Perform requirements review	• FDR pass decision
Product owner Systems engineer	SE PO	340	Finalize the stakeholder needs and system requirements • Baseline the stakeholder needs register • Baseline the system requirements document	• Baselined stakeholder needs register • Baselined system requirements document
Product owner	PO SM SE Development team T&E	350	Manage the product backlog • Break down requirements into workable issues • Assign point values to the work items • Prioritize the product backlog • Perform sprint planning	• Updated product backlog • Populated sprints
Product owner	PO SM SE Development team T&E	360	Perform sprints • Software development • Modeling • Database development • Procurement • Testing	• Demonstration of current product development through a Sprint review
Product owner	PO SE T&E Development team	370	Does the team feel an increment of the system is ready for release and delivery to the stakeholder? • Yes: Hold a publish authorization review • No: Hold a Sprint review	• Publishable increment decision
Product owner Stakeholder	PO SE T&E Development team	380	Sprint review • Demonstration of Sprint work accomplished • Determine if the system is ready for FDR	• Sprint closeout presentation • Record of accepted Sprint items • System ready for FDR determination

(continued)

Table 3.2 (continued)

Individual(s) responsible	SME(s) involved	Step	Action	ARTIFACT
Product owner Stakeholder	PO SE T&E Development team	390	Publish authorization review • Demonstration of Sprint work accomplished • Determination if the system increment is ready for release • Stakeholders sign off on publish authorization form, if publish is authorized	• Sprint closeout presentation • Record of accepted Sprint items • Publish system increment decision • Publish authorization form, potentially • Published system increment, potentially • System ready for FDR determination
Product owner Stakeholder	PO SE	400	Permission to publish the system increment? Yes: Proceed to publish the system increment No: Proceed to system ready for PRR determination	• System increment publish decision
Product owner	PO SE T&E Development team	410	Publish the approved system increment to the production environment • Inform stakeholders of release publication date and potential downtime • Ensure training is updated • Snap a baseline for the published system	• Published system increment • Published system baseline
Product owner Stakeholder	SE PO	420	Is the system ready for PRR? • Yes: Move to production readiness review • No: Proceed to requirements review	• System ready for PRR decision
Product owner Stakeholder	SE PO	430	Is there a change to stakeholder needs or system requirements? • Yes: Move to change management board • No: Proceed to manage product backlog	• Change to stakeholder needs or system requirements decision
Product owner Stakeholder	SE PO	440	Hold a change management board to review proposed changes to the system model, stakeholder needs, or system requirements	• Approved or denied change requests

(continued)

Table 3.2 (continued)

Individual(s) responsible	SME(s) involved	Step	Action	ARTIFACT
Product owner Stakeholder	SE PO	450	Are there changes to the stakeholder needs, system requirements, or system model? • Yes: Update stakeholder needs, system requirements, and the system model, as necessary • No: Proceed to manage product backlog	• Change approval decision
Product owner Stakeholder	SE PO	460	Update the stakeholder needs, system requirements, and system models as appropriate	• Updated stakeholder needs registry • Updated system requirements document • Updated system models
Product owner Stakeholder	SE PO TPM	470	Decision point: PRR At the PRR, the stakeholder and product owner decide whether the system is ready for final product delivery Question to answer: Is the system sufficiently complete that stakeholder needs are met satisfactorily enough to finalize the project and move to sustainment? NOTE: This does not mean that all requirements have been developed or that the product backlog is empty. This means that the stakeholder is satisfied with the current state of the system as a working viable product. Remaining requirements and work items will be triaged and packaged to be transferred for sustainment efforts	• Transferrable product backlog
Product owner Stakeholder	SE PO TPM	480	PRR passed? Yes/pass: Transition to finalize product documents No/fail: Proceed to changes to requirements?	• PRR pass decision

(continued)

Table 3.2 (continued)

Individual(s) responsible	SME(s) involved	Step	Action	ARTIFACT
Product owner	SE PO TPM	490	Finalize project and product documents for transfer to sustainment activity and project archival • Finalize stakeholder needs registry • Finalize system requirements document • Finalize system models • Project artifacts ready for closeout • Create gold discs, if applicable • Prepare and finalize system training documents for delivery • Transfer any remaining work items in the product backlog into change requests or task items with stakeholder input	• Transferable product backlog • Outstanding requirements as documented in the system requirements document • Outstanding stakeholder needs as documented in the stakeholder needs registry • Project artifacts ready for closeout • Gold discs for delivery • Finalized system training documents
Product owner	SE PO	500	Perform the final system deployment • Deploy software to production servers • Deliver gold discs to customers • Perform final system training	• Deployed system and associated artifacts
Product owner	PO SE	510	Transition artifacts and documentation to sustainment activity	• Transferred artifacts • Transferred documents
Product owner	PO TPM	520	Perform project closeout activities. • Closeout project plan • Finalize financial documents • Archive project documents	• Final project plan • Final financial documents

and application of systems engineering methods. All projects had various levels of implementation of Agile or Predictive project management, as well as varying levels of requirements management. The application of the SEFAD process is

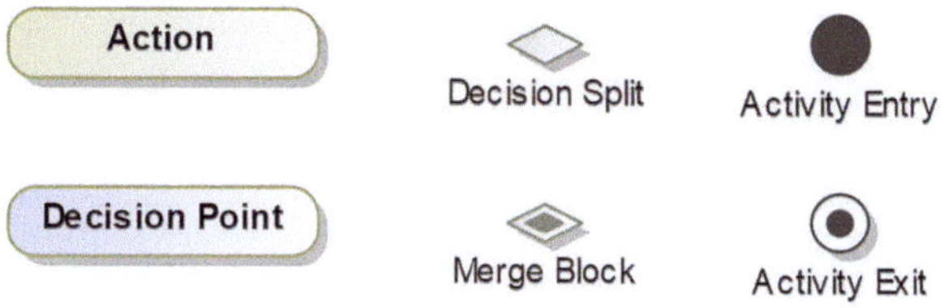

Fig. 3.4 SEFAD process key

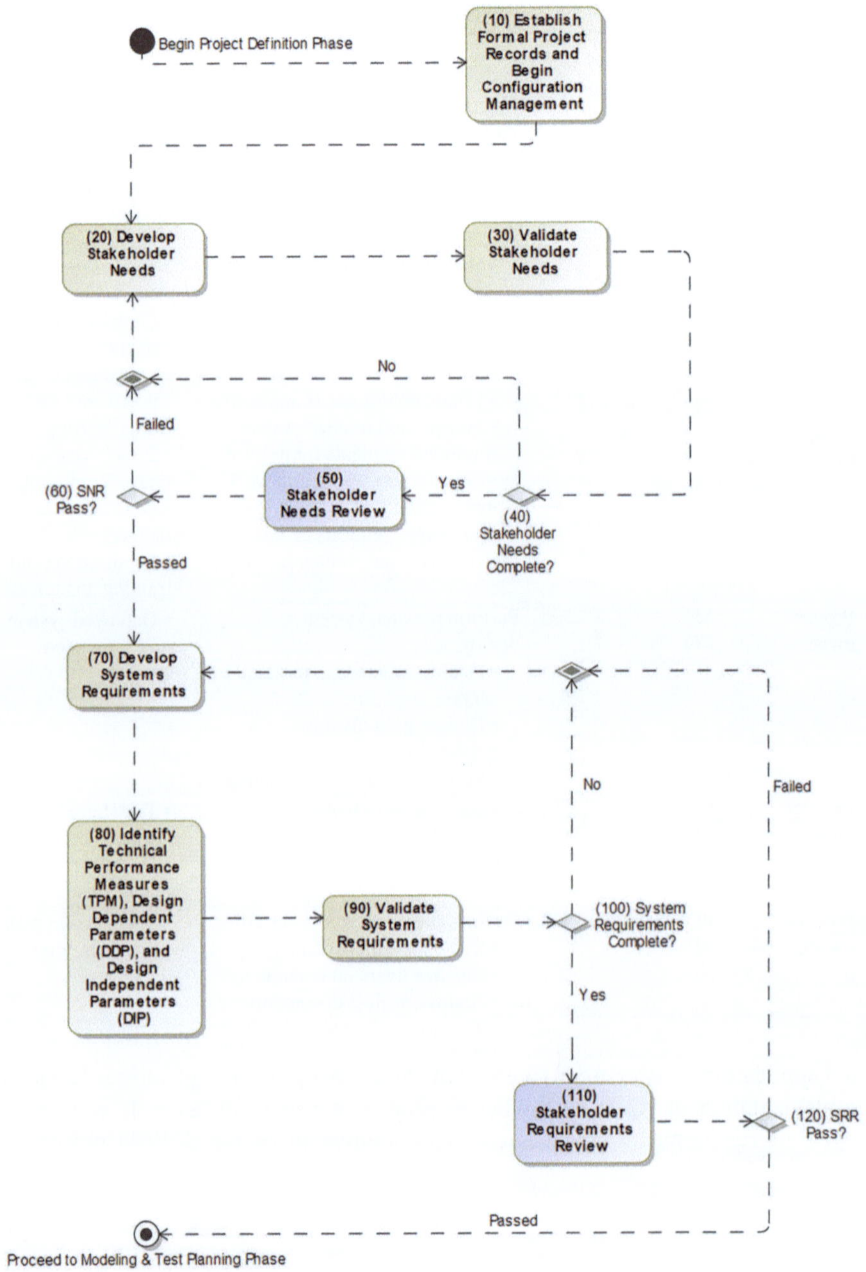

Fig. 3.5 SEFAD project definition phase

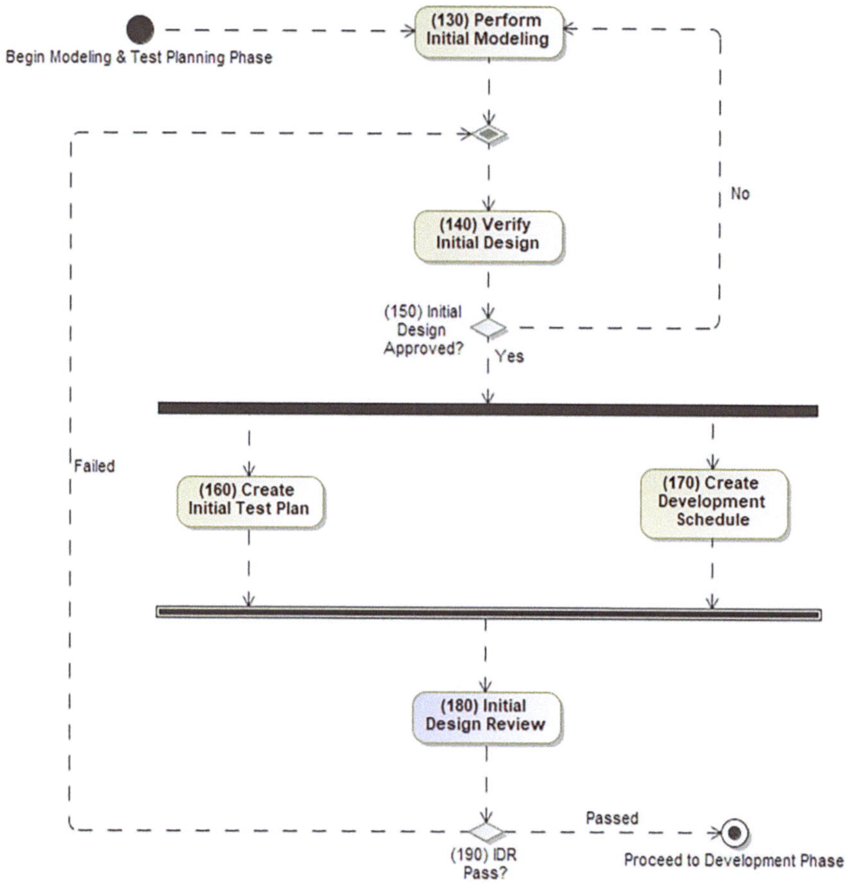

Fig. 3.6 SEFAD modeling and test planning phase

proven to provide increased systems engineering rigor and benefits to cost, schedule, and performance metrics.

Specific DoD project implementation, as of time of writing:

- The DLMS project has implemented the following:
  - Documented and mature Agile Scrum project execution.
  - Customer driven rigorous requirements management.
  - Requirements baseline of existing system ("As built").
  - Wireframing and light SYSML usage.
  - Customer driven change management.
  - SEFAD process implementation.
- The DPAS project has implemented the following:
  - Documented and growing Agile Scrum project execution.
  - Customer driven rigorous requirements management.

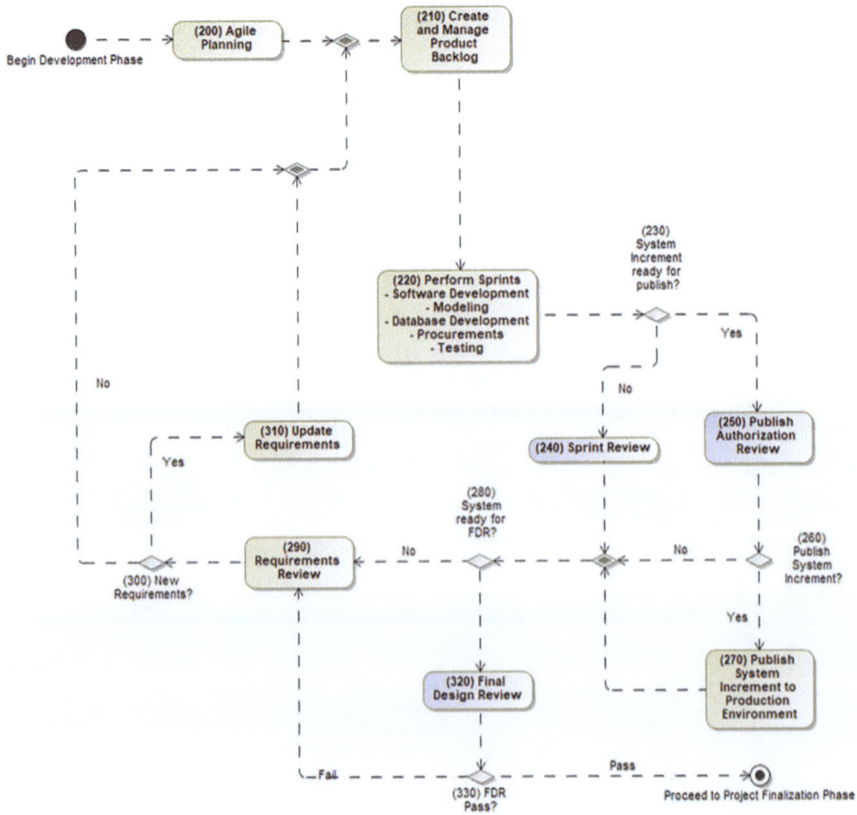

Fig. 3.7 SEFAD development phase

- Positive progress toward requirements baseline of existing system ("As built").
- Customer driven change management.
- SEFAD process implementation.

- The DWMS project has implemented the following:
  - Documented and growing Agile Scrum project execution.
  - Customer driven rigorous requirements management.
  - Positive progress toward requirements baseline of existing system ("As built").
  - Customer driven change management.
  - SEFAD process implementation.
  - Cross-department partnership in systems engineering focused on Agile development.
- The DoD Software Hosting System (DSHS) project has implemented the following:

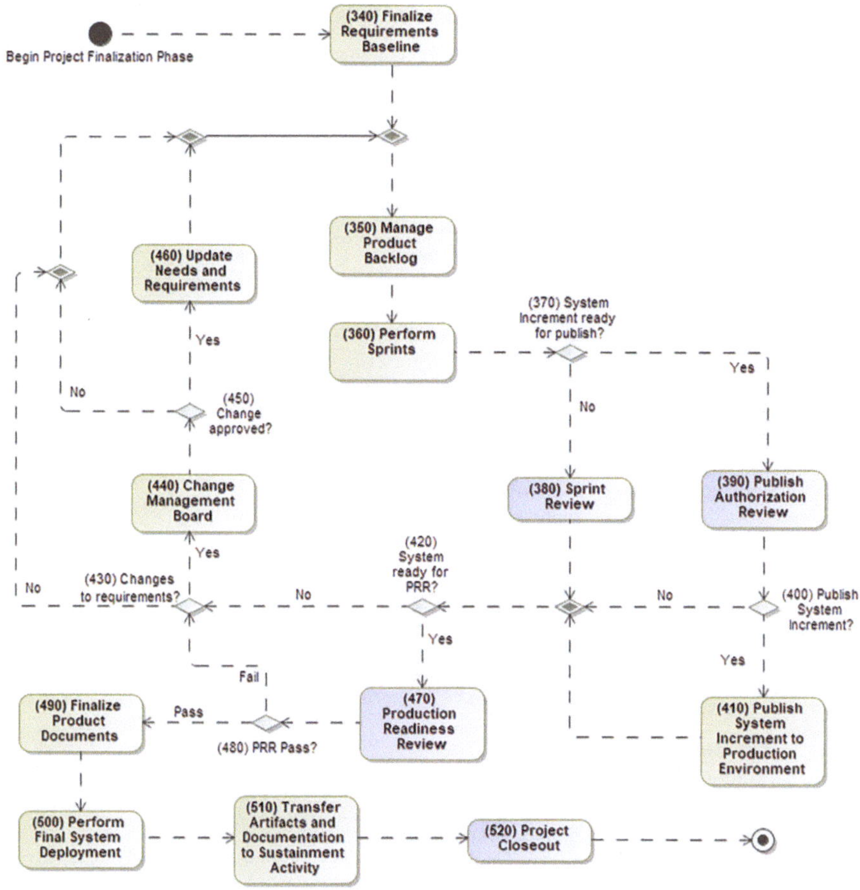

Fig. 3.8 SEFAD project finalization phase

  - Customer driven rigorous requirements management.
  - Positive progress toward requirements baseline of existing system ("As built").
  - SEFAD process implementation.

- The DESS project has implemented the following:
  - Documented and growing Agile Scrum project execution.
  - Customer driven rigorous requirements management.
  - Completion of a requirements baseline of existing system ("As built").
  - SEFAD process implementation.

Office of the Undersecretary of Defense (OUSD) DMSMS Program Office implementation:

- The Enterprise Part Management System project has implemented the following:

Table 3.3 SEFAD performance assurance and audit checklist

Item	Indication of process compliance	Artifacts of process compliance			
		Record	Location	Custodian	Retention
1.	Stakeholder needs register	Stakeholder needs register, needs baselines	See configuration management plan for location/responsibility/retention policies		
2.	System requirements document	System requirements document, requirements baselines			
3.	Configuration management plan	CM plan			
4.	Project plan	Project plan documents			
5.	Stakeholder needs review (SNR)	Entrance/exit criteria, approval authority signoff			
6.	System requirements review (SRR)	Entrance/exit criteria, approval authority signoff			
7.	System models	Documented models			
8.	Test plan (unit, integration, verification)	Documented test plan			
9.	Test matrix	Documented test matrix			
10.	Test procedures (unit, integration, verification)	Documented test procedures			
11.	Initial design review (IDR)	Entrance/exit criteria, approval authority signoff			
12.	Project schedule (FY, sprints, prototypes, delivery)	Project schedule			
13.	Product backlog	Work item tracking system (JIRA, GitLab, etc.)			
14.	Sprint logs	Work item tracking system (JIRA, GitLab, etc.)			
15.	Sprint test results	Test records			
16.	Final design review (FDR)	Entrance/exit criteria, approval authority signoff			
17.	Project plan	Approved project plan			
18.	Lessons learned	Documented lessons learned			

Note: This table is intended to be used by internal quality auditors, managers, and process practitioners to monitor and demonstrate adequate performance of this process

- Documented and growing Agile Scrum project execution.
- Documented and growing Scaled Agile Framework execution.
- Customer driven rigorous requirements management.
- Completion of a requirement baseline prior to project execution.
- SEFAD process implementation.
- Rigorous and thorough use of Model-Based Systems Engineering.
- Systems modeling using SySML.
- Department of Defense Architectural Framework Views and Models.

Other government branches or agencies implementing this methodology:

- Puget Sound Naval Shipyard (PSNS) Intermediate Maintenance Facility (IMF), Code 300.1.
- Naval Undersea Warfare Center Division Keyport, Undersea Weapons Department, Engineering and Production Enablement.
- Naval Undersea Warfare Center Division Keyport, Fleet Readiness Department, Electrical Engineering Applied Technology Branch.
- Naval Undersea Warfare Center Division Newport, In Service Engineering Activity 1533.
- Naval Surface Warfare Center (NSWC) Crane Division, Microelectronics Assurance Branch (GXVR).
- Bureau of Labor Statistics, Office of Technology and Survey Processing.
- Defense Human Resources Activity, Defense Manpower Data Center, Technical Services Division.

3.3.4 SEFAD: Conclusions

The ability for Agile Scrum practitioners to swiftly and decisively pivot to changes in customer needs, goals, and requirements is key to the success of these types of projects. Predictive style projects are far too rigid for software development and lose this ability to be flexible with desired customer changes. The use of User Stories in Agile Scrum projects allows for rapid refocus of software development projects but comes with the drawback of documenting requirements in a very abstract manner, much more akin to goals than requirements. The use of thorough, standardized, and atomic requirements in the form of systems engineering shall statements increases the ability for developers and customers to communicate with one another, directly leading to improved system and problem domain understanding. Furthermore, clear requirements definition and management increases the ability for the development team to perform Verification and Validation activities. Additionally, performing modeling activities allows developers to better communicate project vision to customers, receiving rapid feedback, lowering the probability of misunderstanding of the system and problem domain, and decreasing the risk of defect generation and schedule slippage. Therefore, a process that thoughtfully implements systems engineering rigor, in the form of requirements management, modeling, and V&V, while retaining the flexibility of Agile Scrum incremental development and close customer collaboration, is needed.

The SEFAD process uses a "rigid—flexible—rigid" method of developing software. When considering the Systems Engineering V, the upper left-hand portion of the V represents project initiation. In Predictive style projects, this is where requirements are generated and validated with customers, along with initial project management activities. In phase 1 of the SEFAD process, this is still true. Customers and developers work closely to create a needs, goals, and requirements baseline to establish initial system and problem domain understanding. Modeling and V&V activities also occur on the left-hand side of the Systems Engineering V, which is represented in phase 2 of the SEFAD process. Both of these activities are considered "rigid" or standardized, fully documented, and following a prescribed path. When transitioning into development activities, the SEFAD process differs from the Systems Engineering V execution in that everything becomes "flexible," or Agile, in that all project artifacts can be changed, updated, added to, and deleted, including requirements. This allows customers to refocus on a project without having to restart the process. The differentiation of the SEFAD approach to pure Agile Scrum is an addition of scheduled requirements meetings with customers. Needs, goals, and requirements are traced directly to User Stories, and requirements repository for the project is maintained. Post Sprint Review, customers are immediately asked to communicate any desired additions or deletions in requirements, as well as reprioritization of the requirements repository and Product Backlog. If new requirements are generated, they are traced to existing User Stories or new User Stories are generated and placed in the Product Backlog. In this way, the flexibility of Agile Scrum is maintained while also leveraging the clarity of systems engineering requirements management. When the customer is satisfied with the project and has identified an end state, the team then transitions into phase 4, baselining all system and project artifacts and minimizing changes. This represents a transition back into a rigid methodology, increasing the ability of the team to deliver on schedule. Through the use of a "rigid-flexible-rigid" method to develop software, increases in system understanding and documentation are gained while lowering the risk of negative impacts to cost schedule and performance.

3.4 Discussion Questions

Defining criticality: How does the text define "criticality," and what factors determine whether a system should be classified as critical? Can criticality exist along a spectrum rather than as a binary classification?

System of systems risk: How can a defect in a non-critical subsystem propagate within a larger System of Systems architecture to create critical, emergent failures? What design or verification practices help mitigate this risk?

Agile in critical contexts: The text warns that pure Agile practices can undermine safety, security, and mission assurance. Which specific Agile principles are most at odds with critical system development, and how might they be adapted without losing agility?

Safety and mission assurance: What does it mean to integrate safety and mission assurance into each development increment? How can fault containment and rigorous validation be embedded in an iterative Agile lifecycle?

Traceability and requirements management: Why is bidirectional traceability essential in critical system development, and what failures can result when Agile teams rely solely on Epics, Features, and User Stories without formal traceability structures?

Governance and compliance: How can Agile development teams reconcile the need for flexible iteration with the fixed regulatory and certification demands of frameworks such as DoD 5000, FAA DO-178C, or NASA NPRs?

Cultural divide: This chapter describes a deep cultural difference between traditional engineers and Agile practitioners. What are the root causes of this divide, and how can leadership foster mutual trust and integration between these groups?

Risk management and fault tolerance: How does Agile's focus on adaptability complicate traditional fault tolerance strategies such as redundancy analysis, hazard modeling, and failure containment? What hybrid practices could preserve both flexibility and safety?

Documentation and knowledge capture: Why does the Agile principle of "working software over documentation" present particular danger in critical domains? How can teams build living documentation that supports sustainment and certification without impeding velocity?

The case for SEFAD: The SEFAD model proposes a "rigid–flexible–rigid" structure to balance agility and rigor. How does this structure address the risks identified earlier in the chapter, and what evidence suggests it can improve cost, schedule, and performance outcomes?

Reference

1. Office of the Under Secretary of Defense for Research and Engineering, *Systems Engineering Guidebook* (Office of the Under Secretary of Defense for Research and Engineering, Washington, 2022)

Chapter 4
Best Practices: Implementing Rigorous Requirements Management into Agile Scrum

4.1 Introduction: The Intersection of Rigor and Agility

Well written requirements are the foundation of all well-engineered systems and successful engineering efforts. They define what must be achieved, how performance is measured, and under what constraints the system must operate. In critical software systems, requirements enforce discipline, accountability, and assurance. A poorly written requirement can cascade through the development lifecycle, generating integration errors, safety hazards, and mission failure. Requirements define the problem domain and, in doing so, directly increase engineering understanding of the system to be built.

In traditional engineering contexts, the Predictive style shall statement dominates. Systems engineers define requirements through exhaustive analysis, allocate them across architecture layers, and establish baselines long before implementation begins. These baselines remain fixed except under formal change control. This approach guarantees consistency, but it also constrains flexibility. Whether due to an ever-evolving mission, emergent and disruptive technologies, realized component and language obsolescence, or through stakeholder feedback, when requirements change, systems managed in a Predictive style can become brittle, slow to respond, and cost-intensive to adapt.

Agile frameworks emerged from software development to address this rigidity. They favor incremental delivery, stakeholder collaboration, and iterative feedback. Instead of waiting for all requirements to stabilize, Agile teams deliver functionality in short Sprints, validating outcomes continuously. However, while this flexibility suits commercial or consumer software, it clashes with the assurance expectations of regulated environments. In defense or medical systems, uncontrolled iteration can erode traceability, obscure verification intent, and violate compliance obligations.

D. Rosson, *Systems Engineering in Agile Software Development*,
https://doi.org/10.1007/978-3-032-14613-7_4

The SEFAD model attempts to reconcile these paradigms. SEFAD integrates the formal rigor of systems engineering, such as requirements decomposition, traceability, and configuration control, with the adaptability of Agile Scrum. Under SEFAD, "shall" statements anchor system integrity and problem domain knowledge, while user stories operationalize them for backlog management and execution. The result is a lifecycle that is both responsive and compliant, capable of evolving without losing architectural discipline or audit readiness.

4.2 Defining Requirements in the Context of SEFAD

4.2.1 The Role of Requirements

A requirement is a contract between intent and realization representing knowledge transfer between a customer and a developer. It captures an obligation that must be verifiably satisfied to achieve a defined outcome. According to INCOSE, a requirement is "a statement that identifies a capability or condition that a system must meet or possess to satisfy a need, contract, standard, or specification [1]". Within SEFAD, requirements function as the architectural DNA of a project, providing the stable reference points around which Agile iteration revolves. They are the seed upon which a system will grow and the roots that maintain it throughout the lifecycle.

4.2.2 The Requirements Hierarchy

Requirements exist at multiple abstraction levels, forming a cascade of intent:

- Stakeholder needs: Human or organizational desires expressed in operational language (e.g., "The fleet needs to track all undersea contacts in real time").
- System functional requirements: Formalized "shall" statements specifying measurable system behavior.
- Derived and Allocated Requirements: Technical constraints propagated down the architecture (e.g., subsystem or software-level obligations).
- Interface requirements: Conditions governing interaction between components or systems.
- Verification requirements: Defined evidence needed to confirm each "shall" is satisfied.

SEFAD links each level through traceable relationships, forming a closed feedback loop that connects stakeholder intent to validated performance.

4.2.3 The Dual Nature of Requirements

In critical software systems development, requirements serve two masters: engineering rigor and development agility. They must be stable enough to guide design and testing, yet flexible enough to evolve with emerging understanding. SEFAD addresses this duality through tiered granularity. Core "shall" statements only become fixed once baselined and affixed to a release version. Until this time, they are linked to user stories, tasks, and acceptance criteria with the capability to evolve freely within each increment. This provides the agility to respond to feedback without undermining system bas4.3 The Attributes of Good Requirements (INCOSE Framework).

4.2.4 The 10 Attributes of Good Requirements

The INCOSE Guide to Writing Requirements (2023) outlines ten attributes of effective requirements. These serve as evaluation criteria for every "shall" statement produced under SEFAD.

- Necessity: Every requirement must represent an essential characteristic of the system. If removing it would not degrade performance, safety, or compliance, it should not exist. Necessity ensures focus on mission-critical elements rather than speculative "nice to haves."
- Implementation Independence: Requirements must describe what the system shall do, not how it will do it. This avoids premature design decisions. Implementation details belong in design artifacts, not in requirements documents.
- Feasibility: Each requirement must be technically, financially, and temporally achievable. Unrealistic or unbounded statements create downstream instability. Feasibility analysis involves early trade studies, risk assessments, and cost-performance modeling.
- Verifiability: Requirements must be testable through one or more verification methods: inspection, analysis, demonstration, or test. Vague or subjective statements ("user-friendly," "easy to maintain") violate this principle unless supported by quantifiable criteria.
- Unambiguity: Ambiguity is the enemy of execution. Requirements must have one and only one interpretation. Controlled vocabulary, consistent units of measure, and standardized structure minimize misinterpretation.
- Completeness: A complete requirement contains all information necessary to understand its intent and constraints. Incomplete requirements invite assumptions and inconsistent implementation.
- Consistency: Requirements must not conflict internally or with external baselines, standards, or contractual documents. SEFAD enforces consistency through peer review and automated trace checks within the digital thread.

- Singularity: Each requirement expresses one concept only. Compound statements such as "The system shall log user activity and generate monthly reports" are split into distinct atomic requirements to enable discrete verification.
- Traceability: Traceability links requirements upward to needs and downward to design, implementation, and test evidence. This bidirectionality allows auditors and engineers to follow cause and effect across the entire system.
- Correctness: A requirement must accurately capture stakeholder intent. Incorrect requirements such as those derived from misinterpreted needs or obsolete standards, are worse than no requirement at all.

4.2.5 Application in SEFAD

In SEFAD, compliance with the INCOSE attributes of good requirements is not treated as a one-time activity or a documentation milestone. Rather, it is a continuous evaluation process embedded within the engineering lifecycle. Every major technical review serves as a formal checkpoint to assess each requirement against the ten INCOSE criteria: necessity, implementation independence, feasibility, verifiability, unambiguity, completeness, consistency, singularity, traceability, and correctness. These reviews ensure that requirements maturity evolves in parallel with system maturity and that the integrity of the baseline strengthens rather than erodes as Agile iterations proceed.

The Stakeholder Needs Review (SNR) represents the first formal application of these attributes. At this stage, emphasis is placed on necessity, correctness, and completeness, as shown in Fig. 4.1. The development team verifies that each stated need corresponds to an actual stakeholder goal and that it has been captured accurately and without assumption. This early evaluation prevents superfluous or misinterpreted needs from propagating downstream into formal system requirements. The goal is to confirm that what is being asked for is both essential and relevant to the mission before translation into "shall" statements occurs.

The System Requirements Review (SRR) extends this assessment to the system-level baseline. Here, the focus shifts toward verifiability, feasibility, and traceability, as shown in Fig. 4.2. Each requirement must be measurable, achievable within program constraints, and linked explicitly to a parent stakeholder need. The SRR marks the formal establishment of the initial requirements baseline, which becomes the governing reference for subsequent design, test, and validation activities. Under SEFAD, SRR is an engineering quality control event that actively tests the coherence and rigor of the requirements set.

By the time the project reaches the Initial Design Review (IDR), attention turns to consistency, implementation independence, and singularity, as shown in Fig. 4.3. At this phase, requirements are examined for logical harmony across subsystems, ensuring that no contradictions exist and that design decisions have not prematurely constrained architectural flexibility. Each requirement is reviewed for atomicity—expressing only a single, testable condition—and for alignment with the broader

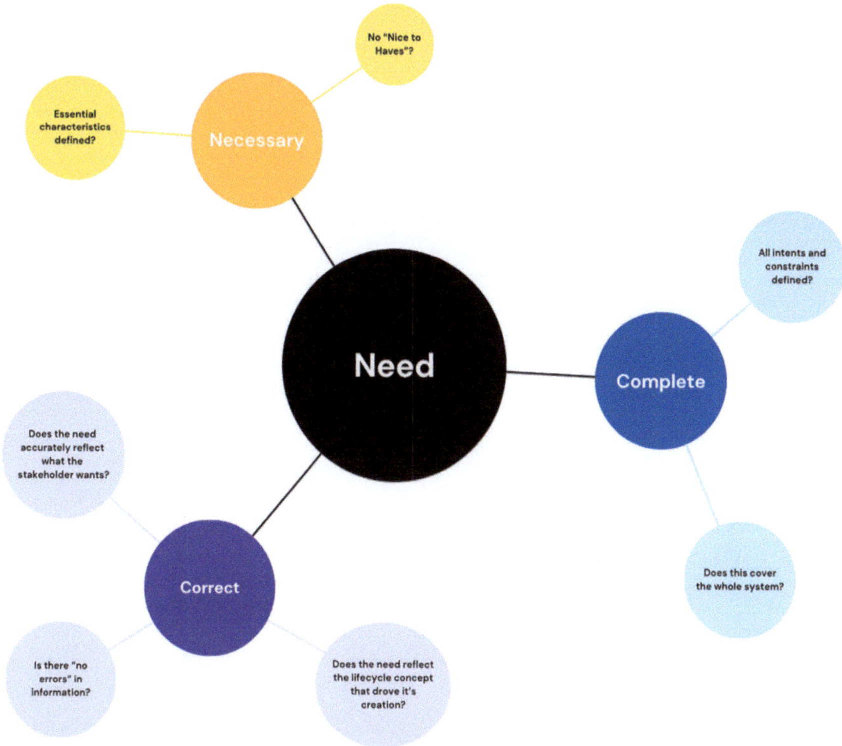

Fig. 4.1 Stakeholder needs attribute map

mission objectives. The IDR thus acts as both a quality filter and a calibration point between design exploration and formal system definition. The test plan is validated against the initial requirements baseline, and the requirements are traced to the first, high-level system model. This creates the all too important foundational traceability to increase system and problem domain knowledge.

What distinguishes SEFAD from traditional processes is that these assessments are not limited to formal milestone events. Agile teams are expected to use the INCOSE attributes as an active checklist during every iteration. Each Sprint review and backlog refinement session includes a short quality check on any new or modified user story, evaluating whether its linked requirement still meets the defined attributes. For example, if a user story introduces new acceptance criteria, the team must confirm that the parent requirement remains verifiable and consistent. This continuous enforcement creates a living culture of rigor—one where compliance is achieved incrementally, not retroactively.

By integrating requirements evaluation directly into the Agile cadence, SEFAD ensures that quality does not accumulate as deferred risk at the end of the project. This represents a major departure from traditional systems engineering models, where requirements audits often occur late in the lifecycle, leading to expensive

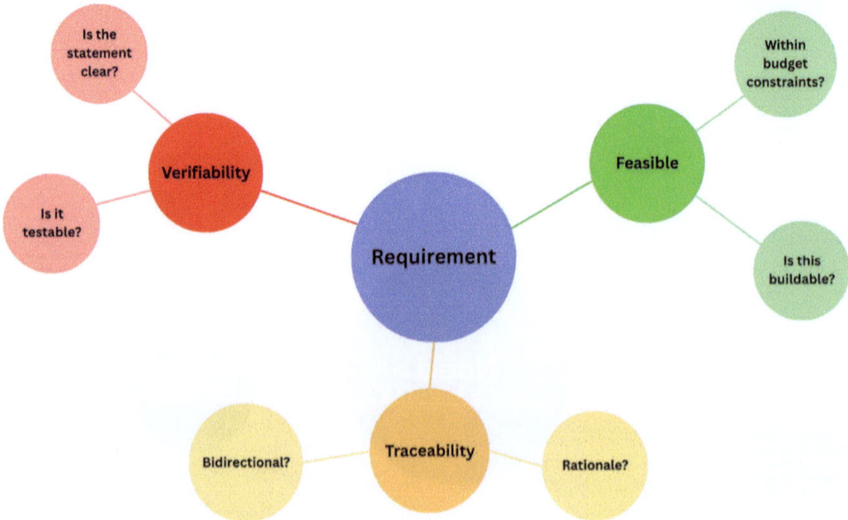

Fig. 4.2 Stakeholder requirements attribute map

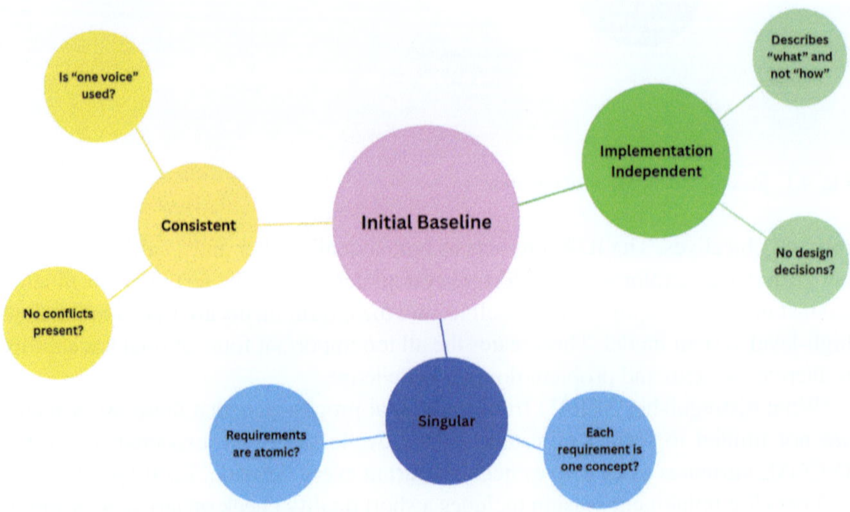

Fig. 4.3 Initial baseline attribute map

rework and schedule slips. Instead, SEFAD institutionalizes real-time requirements assurance, embedding the INCOSE attributes as part of daily practice rather than as an external compliance obligation. The result is a smoother path to design maturity, fewer defects introduced during implementation, and higher stakeholder confidence at each technical review.

Ultimately, this approach transforms the INCOSE checklist from a static compliance document into an operational tool of systems governance. It empowers both engineers and product owners to speak a common language of quality—one grounded in measurable attributes rather than subjective interpretation. Over time, this discipline yields a requirements baseline that is not only auditable but alive: traceable, testable, and continuously validated as the system evolves. Through this, SEFAD bridges the long-standing gap between documentation and execution, ensuring that every increment of Agile progress reinforces, rather than erodes, the foundation of engineering rigor.

4.3 Structuring Requirements for Agile Environments

Traditional requirements management, as practiced within defense and aerospace programs, is grounded in dense specification documents, formalized review gates, and configuration-controlled baselines. Predictive approach assumes stability is achieved through exhaustive up-front definition. Requirements are decomposed, verified, and frozen prior to significant design activity. This approach offers unmatched control and traceability with each "shall" statement documented, validated, and linked to higher-order needs. The process produces tangible assurance artifacts such as System Requirement Documents (SRDs), Interface Control Documents (ICDs), and Requirements Traceability Matrices (RTMs). These deliverables serve as the backbone for compliance audits, verification planning, and contractual oversight. Yet, their strength is also their weakness: heavy documentation can become a barrier to responsiveness, slowing innovation and constraining adaptability when stakeholder needs evolve mid-program.

In contrast, Agile environments approach requirements as dynamic constructs, emphasizing human interaction and iterative understanding over static documentation. User Stories, Epics, and Features replace multi-volume specifications. Instead of long baselining cycles, Agile teams rely on lightweight artifacts such as product backlogs, definition of done acceptance criteria, and Sprint Reviews to convey intent. Collaboration replaces formal traceability and open dialog with stakeholders replaces document review. This structure thrives in fast-moving software domains where learning occurs continuously through feedback, testing, and rapid iteration. However, in mission-critical or safety-critical systems, such informality introduces unacceptable risk. Without the discipline of traceability and controlled baselines, design decisions can become opaque, dependencies can be lost, and verification can be incomplete. Agile's strength in adaptability must therefore be bounded by the rigor of systems engineering discipline.

The SEFAD framework bridges these two worlds by establishing structure with flexibility through a process architecture that fuses Predictive discipline with Agile responsiveness. SEFAD does not discard traditional baselines; instead, it redefines how they evolve. In early phases, the system begins with traditional engineering artifacts such as needs statements, goals, and requirements written in "shall" form and evaluated through formal reviews. These artifacts establish the initial baseline, ensuring clarity, auditability, and alignment with stakeholder intent. As the project transitions into Agile execution, SEFAD maintains that baseline as a living reference rather than a frozen relic. Each User Story is directly linked to one or more of these governing "shall" statements, preserving the chain of traceability even as iterative development unfolds.

Through this linkage, SEFAD creates a dual-layered requirements ecosystem. The Predictive layer preserves compliance, enabling traditional engineering oversight, verification planning, and technical performance management. The Agile layer enables adaptability, translating requirements into actionable increments that can be demonstrated, tested, and refined in near real time. Traceability threads between the two layers ensure that changes in one are immediately reflected in the other. When a Product Owner modifies a User Story or reprioritizes backlog items, the corresponding requirement relationships are updated within the traceability framework, maintaining alignment between system design and contractual obligation. Conversely, when a formal requirement changes due to mission evolution or regulatory feedback, the linked Agile artifacts adapt automatically, allowing the development team to pivot without losing control of scope or intent.

This structure transforms documentation from a static burden into a dynamic control system. Instead of being archived after review, requirements artifacts become operational instruments of governance, continuously referenced, refined, and re-validated. The SEFAD process embeds this linkage into its digital thread through modeling tools, configuration management repositories, and requirements management systems such as IBM DOORS, Jama Connect, Jira R4J, or SysML Requirements Diagrams. These tools provide the backbone for maintaining vertical and horizontal traceability between specifications, user stories, test cases, and design models. As a result, verification evidence, defect logs, and Sprint outputs are all contextually tied to the authoritative source requirement, closing the loop between engineering intent and Agile execution.

The result is a synthesis of Predictive control and Agile flow. The rigor of traditional systems engineering provides the structural integrity necessary for safety, mission assurance, and certification, while the agility of iterative delivery ensures that the system remains relevant, responsive, and continuously validated by stakeholders. SEFAD's bridging approach allows documentation and dialog to coexist—not as adversaries, but as complementary mechanisms of assurance and learning. The dense specification and the lightweight backlog no longer represent competing philosophies; they are instead connected nodes within a single, traceable system of understanding. In this way, SEFAD transforms the perceived dichotomy between rigor and flexibility into a productive continuum of controlled adaptability, uniting the best attributes of both paradigms into a coherent methodology for modern systems engineering.

4.3.1 The "Shall" Statement Template

A properly constructed requirement in SEFAD follows the standard INCOSE syntax:

The [system/subsystem/component] shall [action verb] [object] [quantitative measure] [rationale, if applicable].

Example:

The Coffee Maker Brewing System shall heat water to a temperature between 90 °C and 96 °C at the point of infusion with the coffee grounds.

These shall statements are stored in requirements repositories utilizing tables or matrixes. Each requirement is numbered and placed hierarchically, in parent-child relationships, thereby creating natural traceability. An example of this is shown in Table 4.1.

4.3.2 The User Story Template

User stories capture functional intent in the customer's language. SEFAD extends the Agile Connextra format by linking stories directly to "shall" statements:

As a [user role], I want [capability] so that [benefit].

Linked Requirement: [SR-ID]

Acceptance Criteria: [Quantitative or demonstrable test conditions].

Example:

As a coffee machine operator, I want the coffee maker to heat the brewing water to the optimal temperature range, so that my coffee has consistent flavor, aroma, and strength every time I brew it.

Linked Requirement(s): SR-023

Acceptance Criteria:

- When brewing begins, the system heats the water to a temperature between 90 °C and 96 °C at the point of contact with the coffee grounds.
- The temperature remains within this range for the duration of the brewing cycle.
- The system automatically adjusts heating as needed to maintain this temperature range under normal environmental conditions (ambient 20–25 °C).
- A test brew demonstrates that the average temperature at infusion meets the specification in three consecutive trials using calibrated instruments.

This ensures that every Agile deliverable has a traceable root in formal requirements, preserving auditability.

Table 4.1 Coffee machine requirements table

Prioritization	Source of requirement	Requirement #	Requirement type	Requirement level	Requirement statement	Rationale	Validation activity	Verification activity
Core	BN.3, BI.7, IEC 60335–2-6, BR.3	CM.SR.1	System	Functional	The CM2 shall meet the international appliance testing and certification standard of IEC 60335–2-6 to ensure that the appliance can be sold internationally.	By engineering the CM2's electrical system to work in multiple international countries, efforts will be saved in future re-engineering if marketing decides to move to other countries.	Demonstration	Inspection
Core	BN.3, BI.7	CM.SR.1.1	System	Functional	The CM2 shall operate on 110 V so that it can be used in the US	By engineering the CM2 to work both with 110v, the machine can operate on the united state electrical grid	Demonstration	Inspection
Core	BN.3, BI.7	CM.SR.1.2	System	Functional	The CM2 shall operate on 220 V so that it can be used in Europe	By engineering the CM2 to work both with 220 V, European electrical grid	Demonstration	Inspection

Prioritization	Source of requirement	Requirement #	Requirement type	Requirement level	Requirement statement	Rationale	Validation activity	Verification activity
Core	BN.3, BI.7	CM.SR.1.3	System	Functional	The electrical cord for the CM2 shall detach so that users can replace the cable with a different one to allow for various types of electrical sockets to be used for power	By engineering the CM2 with a removable and replaceable electrical plug, the applicance will be able to be easily customized to individual country markets as not all countries use the same style of electrical plug	Demonstration	Inspection
Essential	CN.4, CN.13, EnergyStar commercial coffee maker Cerfification guidance	CM.SR.2	System	Functional	The coffee maker shall operate on at least 30% less energy than standard small batch type II coffee makers as specified by the EnergyStar commercial coffee maker certification guide to ensure EnergyStar certification	To meet EnergyStar standards, the coffee maker must be engineered to utilize at least 30% less energy than standard models of the small batch type II coffee maker classification	Demonstration	Inspection
Core	CI.10, CI.12, CI.13, CN.1, CN.13	CM.SR.3	System	Functional	The CM2 shall grind beans upon command	Customers desire coffee that is ground right before brewing begins, as the flavor is better. They also desire ease of use and a self contained system	Demonstration	Inspection

(continued)

Table 4.1 (continued)

Prioritization	Source of requirement	Requirement #	Requirement type	Requirement level	Requirement statement	Rationale	Validation activity	Verification activity
Essential	CN.4, CN.13, EnergyStar commercial coffee maker Cerfification guidance	CM.SR.3.1	System	Functional	The CM2 shall grind beans using a burr grinder	Customers desire burr grinders as there is a general belief that they are superior style of coffee grinder	Demonstration	Inspection
Core	CI.4	CM.SR.4.1	System	Functional	The CM2 shall brew espresso coffee so that users can drink espresso drinks	Customers desire the ability to brew drip coffee	Demonstration	Inspection
Core	CI.7, CN.1, CN.2	CM.SR.4.2	System	Functional	The CM2 shall brew drip coffee so that users can drink drip coffee	Customers desire the ability to brew espresso coffee	Demonstration	Inspection

4.4 Traceability and Digital Thread Integration

4.4.1 Concept of Traceability

Traceability ensures that no requirement exists without justification, and no implementation occurs without purpose. In software, each function developed can be traced directly to a User Story which in turn can be traced directly to a requirement. In SEFAD, this is enforced through an end-to-end digital thread, connecting:

Stakeholder Needs → System Requirements → User Stories → Tasks → Code Commits →.

Test Results → Verification Evidence.

4.4.2 Example Traceability Record (Text Representation)

Requirement ID: SR-023.

Statement: *The Coffee Maker Brewing System shall heat water to a temperature between 90 °C and 96 °C at the point of infusion with the coffee grounds.*

Linked User Story: US-013 - *As a coffee machine operator, I want the coffee maker to heat the brewing water to the optimal temperature range, so that my coffee has consistent flavor, aroma, and strength every time I brew it.*

Verification Method: Test/Analysis.

Status: Verified, Sprint 7.

Trace To: CM Brew System CM-BREW-01, Test Case CM-BREW-19.

This textual form mirrors the structure of a formal Requirements Traceability Matrix (RTM) or Verification Cross-Reference Matrix (VCRM) and is maintained dynamically within requirements management tools (e.g., JIRA, DOORS Next, etc.).

4.5 The Rigid: Flexible—Rigid Lifecycle Model

The SEFAD lifecycle is intentionally designed to oscillate between two complementary states: strict control and adaptive execution. This duality embodies the balance between the disciplined predictability of systems engineering and the responsive flexibility of Agile. Each state serves a distinct function within the program's evolution. The controlled phases establish structure, verification, and compliance. The adaptive phases deliver functionality, learning, and innovation. When sequenced properly, these modes form a closed feedback system that transforms uncertainty into validated capability. Loops between the phases are introduced to allow for further flexibility in execution and continuous learning, ensuring the Three Ways of DevSecOps are integrated into the lifecycle: flow, feedback loops, and continuous learning.

4.5.1 *Phase I: Project Definition Phase (Rigid)*

Initial stakeholder needs and system requirements are generated, validated, and baselined. The focus is on correctness, completeness, and traceability.

Activities include:

- Elicitation workshops with stakeholders and development engineers
- Development of the Stakeholder Needs Document (SND)
- Drafting of the System Requirements Document (SRD)
- Establishment of verification methods and preliminary test planning
- Approval through the SRR

4.5.2 *Phase II: Modeling and Test Planning (Rigid)*

Initial modeling, design verification, and development schedule are developed. Test planning commences. The focus is on high-level problem domain knowledge transfer and understanding as well as developing a good verification and validation structure.

Activities include:

- Modeling the system at an operational level
- Creation of the initial test plan
- Creation of the development schedule
- Approval through the IDR

4.5.3 *Phase III: Development Phase (Flexible)*

Agile Sprints commence, implementing requirements through user stories. During this phase:

- Each user story is linked to one or more system requirements.
- Sprint Reviews validate partial fulfillment of requirements.
- Emerging insights may generate proposed changes to requirements, requirement priorities, or the addition or subtraction of requirements from the repository.
- Release authorization through Publish Authorization Reviews.
- Transition to Project Finalization through gateway control via the FDR.

4.5.4 Phase IV: Project Finalization (Rigid with Flexible Development)

Once the system reaches functional maturity, the focus shifts to closure:

- Requirements are re-baselined to reflect final implementation.
- All verification evidence is collected and linked.
- Traceability matrices are finalized for audit.
- Requirements changes are more difficult and must be approved through a CCB.
- Project finalized through a production readiness review (PRR).
- Perform handover to sustainment activities.

This lifecycle enforces disciplined agility within boundaries of rigor.

4.6 Verification and Validation Alignment

Verification and Validation (V&V) represent the proof that the system meets stakeholder requirements and the highest expression of engineering rigor within the SEFAD framework. They are continuous processes that permeate every stage of the system's lifecycle. In classical systems engineering, V&V occurs primarily during the right-hand side of the V-model and only after design and implementation are complete. In SEFAD, this boundary dissolves. Verification and validation become continuous assurance mechanisms, interwoven into Agile iteration, ensuring that every increment of delivered capability is not only functional but also compliant, traceable, and mission ready. Figure 4.4 shows how a well-engineered system is both verified, or that the system has good engineering rigor in testing to ensure that requirements are being met and it is defect free, and also validated, or activities have ensured that the system being built is what the customer actually needs.

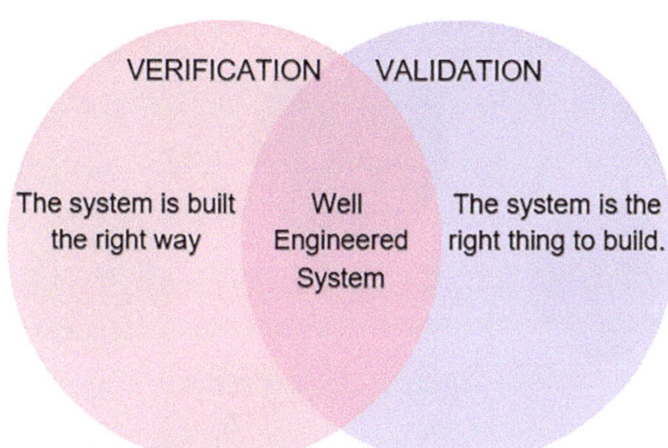

Fig. 4.4 Verification and validation overlap

4.6.1 Verification (Fig. 4.5)

Each requirement is associated with a defined verification method, selected from the standard DoD taxonomy:

- Inspection: Physical or visual examination of documentation or hardware.
- Analysis: Mathematical, logical, or simulation-based evaluation.
- Demonstration: Functional performance shown under controlled conditions.
- Test: Instrumented measurement against quantitative metrics.

4.6.2 Validation (Fig. 4.6)

Validation ensures that the verified system fulfills stakeholder intent. In Agile contexts, validation occurs incrementally through Sprint Review demonstrations and operational assessments of releases. Through the demonstration of incremental builds, Sprint Teams get rapid feedback from customers on solution completeness and whether they are developing the "right thing". This allows them to quickly course-correct if it is found that there is a misunderstanding in stakeholder versus developer vision. Each Sprint delivers partial capability that can be evaluated for mission relevance and not merely technical correctness.

Common questions to be answered during Validation:

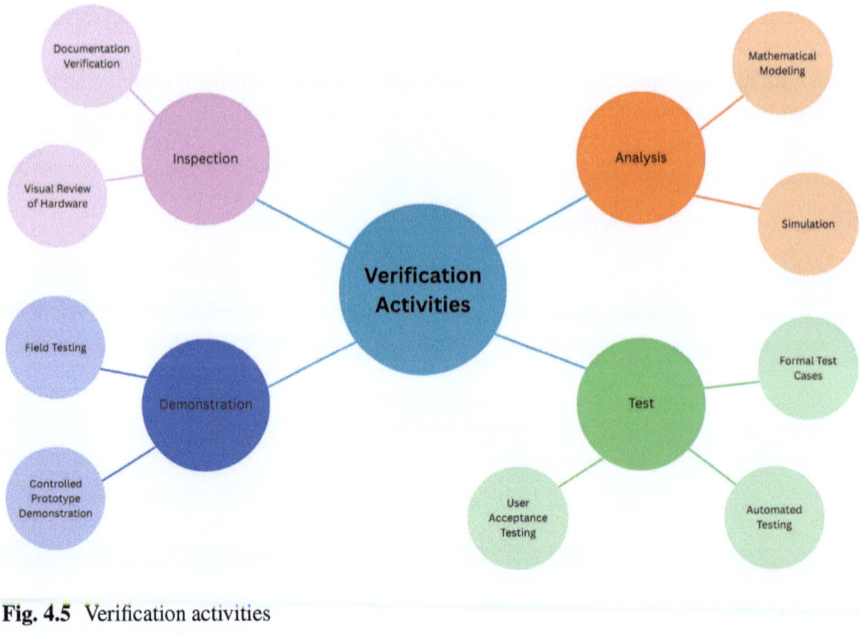

Fig. 4.5 Verification activities

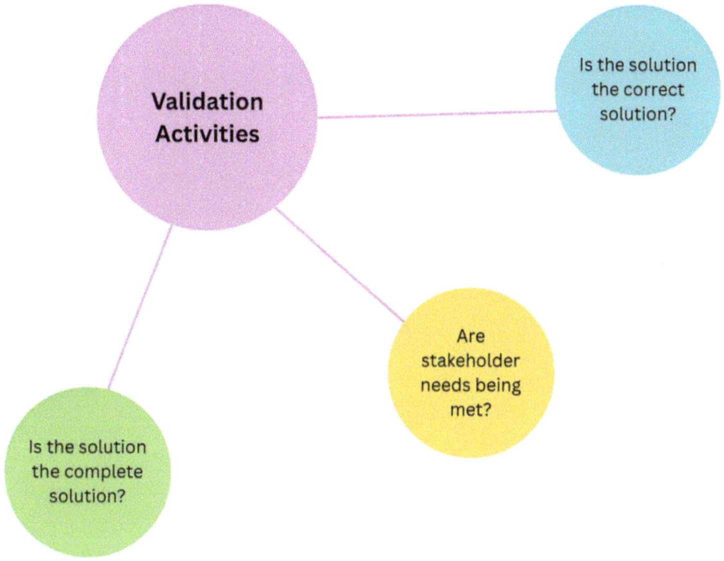

Fig. 4.6 Validation questions

- Is the solution the correct solution to meet stakeholder needs?
- Are all stakeholders' needs being met by the solution?
- Is the solution the complete solution?

4.6.3 The Continuous V&V Loop

SEFAD embeds V&V into every Sprint:

1. User story defines acceptance criteria through mini-validation intent.
2. Development and automated tests verify functionality.
3. Sprint Review provides operational validation opportunity.
4. Formal test artifacts update the test matrix.

This continuous feedback cycle maintains both engineering assurance and customer satisfaction. It is the core of the Three Ways of DevSecOps,[1] ensuring that work is reviewed by customers rapidly after development, eliciting and documenting customer feedback, and updating needs, requirements, vision, and expectations based upon the feedback through continuous improvement and learning. In this way, Agile projects are able to stay ahead of potential changes and adapt processes

[1] For more information and depth on the Three Ways in DevSecOps, please see other works such as "The Phoenix Project" by Gene Kim, George Spafford, and Kevin Behr or "The DevOps Handbook" by Gene Kim, Patrick Debois, John Willis, and Jez Humble.

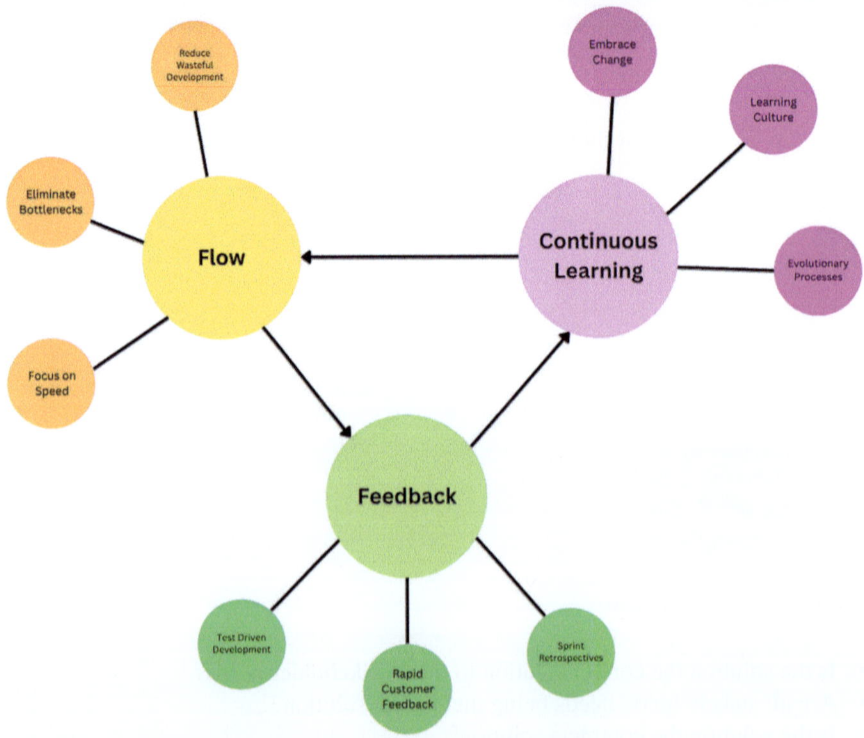

Fig. 4.7 The three ways of DevSecOps in software development

and direction to lower the risk of developing functionality that the customer doesn't need while focusing efforts on what they customer does need. This concept is demonstrated in Fig. 4.7.

4.7 Configuration Management and Change Control Through Tiered Governance

Agile thrives on change, but critical systems require control. This tension lies at the heart of every modern engineering organization that seeks to integrate iterative delivery with mission assurance. The SEFAD model resolves this tension through a structured, yet adaptive approach known as tiered governance. This governance framework is a modern mechanism that enables agility to operate safely within environments that demand accountability, traceability, and compliance, thereby retaining flexibility while ensuring engineering rigor.

4.7.1 The Dual Imperative: Change and Control

Agile principles emphasize flexibility and responsiveness to evolving needs. Requirements are expected to change, and iteration is celebrated as a natural process of discovery. In contrast, the systems engineering discipline, particularly within critical systems, demands controlled evolution. Every modification must be understood, justified, verified, and approved. These two imperatives appear to be at odds. Agile seeks movement and adaptability, always with forward momentum, while systems engineering seeks stability and control. Yet both are essential.

Without control, Agile programs can drift into chaos, accumulating technical debt and compliance risk. Without adaptability, systems engineering becomes too rigid, unable to respond to emergent threats, technologies, or mission changes. SEFAD does not attempt to average these philosophies or compromise between them. Instead, it organizes them into governance tiers that assign decision authority, documentation requirements, and review cadence in proportion to technical and operational criticality. In essence, SEFAD allows rapid change where it is safe, and enforces strict control where it is necessary.

4.7.2 The Structure of Tiered Governance

Tiered governance in SEFAD consists of three interdependent levels: Strategic Governance, Programmatic Governance, and Iterative Governance (Fig. 4.8). Each level has distinct responsibilities, but all share the same objective: to maintain a dynamic equilibrium between responsiveness and rigor.

1. Strategic Governance operates at the enterprise or portfolio level. It defines the mission context, establishes compliance baselines, and sets the engineering policies that govern all subordinate programs. This includes the enforcement of DoD directives such as DoDI 5000.02, DoDI 5000.97, and the DoDI 5000.88. For industry companies, this can be reflected in the enforcement of industry standards such as the INCOSE Systems Engineering Handbook standards for technical process execution or the DO-178C Software Considerations in Airborne Systems and Equipment Certification. Strategic governance ensures that all Agile activities align with overarching system-of-systems architectures, cybersecurity postures, and certification requirements. Change at this level is infrequent and subject to formal Configuration Control Board (CCB) approval.
2. Programmatic Governance exists at the program or project level and manages the implementation of SEFAD processes within a specific domain or capability. It determines how requirements flow down, how traceability is maintained, and how verification evidence is collected. Programmatic governance synchronizes Agile increments with systems engineering technical reviews, such as the System Requirements Review (SRR), Preliminary Design Review (PDR), and Critical Design Review (CDR), ensuring that iterative development cycles feed into for-

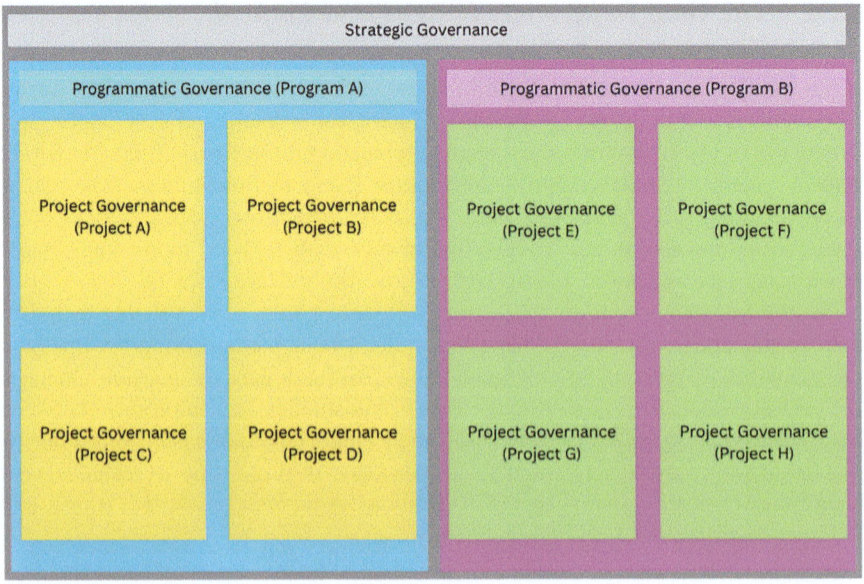

Fig. 4.8 Strategic governance

mal decision gates without losing flexibility. It also maintains the risk register, program schedule, and Earned Value Management (EVM) alignment required for high-accountability environments.

3. Project Governance functions at the Sprint or team level. It defines how Agile ceremonies like the Backlog Refinement, Sprint Planning, daily Scrums, and Sprint Retrospectives, integrate with engineering artifacts and documentation requirements. At this level, governance is lightweight and highly adaptive. Teams can pivot quickly based on stakeholder feedback, if changes are properly linked to traceable requirements and validated through defined acceptance criteria. Project governance ensures that agility occurs within the constraints of engineering discipline rather than outside it.

Together, these tiers form a hierarchical control system that preserves agility where uncertainty is high and enforces discipline where risk is unacceptable. The tiers interact continuously: iterative insights flow upward as feedback to programmatic decision makers, while strategic constraints flow downward as boundaries for execution. The result is not a command-and-control model but a closed-loop governance architecture, similar in spirit to control theory: dynamic, responsive, and self-correcting.

4.7.3 Governance as an Enabler of Agility

Governance in SEFAD is not viewed as an impediment to Agile development. It is an enabler. In traditional Agile frameworks, governance is often treated as an external compliance obligation, added after development. SEFAD internalizes governance as a core operational mechanism. Each governance tier establishes explicit "freedom-to-operate" boundaries. Within those boundaries, Agile teams are free to innovate, experiment, and deliver iteratively. Outside those boundaries, formal systems engineering processes are triggered to ensure that safety, security, or mission integrity are not compromised.

This structure allows SEFAD organizations to scale Agile principles without sacrificing oversight. For example, a Sprint team can refactor a subsystem interface or modify an algorithm without escalating to the program board, as long as the change does not alter verified performance baselines or introduce new external dependencies. However, if that change impacts certification artifacts, cybersecurity controls, or mission assurance parameters, governance automatically elevates the decision to the appropriate tier for review. This conditional escalation model allows for fast execution at the edge and deliberate control at the core.

4.7.4 Tiered Governance and Decision Authority

One of the key innovations in SEFAD's governance model is the explicit mapping of decision authority to each tier. In many organizations, ambiguity about who has authority to approve changes creates friction and delay. SEFAD removes this uncertainty by assigning clear ownership of technical, programmatic, and operational decisions.

- Technical decisions, such as architecture adjustments or interface control updates, belong to the engineering leads within the project governance tier but require traceability updates under programmatic governance.
- Programmatic decisions, including scope definition and risk disposition, rest with the program manager, which ensures that decisions align with system baselines and contractual obligations.
- Strategic decisions, such as architecture evolution or compliance interpretation, are made by the enterprise governance body, typically led by Chief Engineers, Technical Fellows, Directors, and C-Suite Executives.

This delegation of authority creates both autonomy and accountability. Teams can make decisions quickly within their tier but are obligated to maintain transparent documentation and traceability to support higher-tier oversight. In effect, governance in SEFAD distributes power responsibly, ensuring agility does not devolve into anarchy.

4.7.5 Integration with Technical Reviews

Tiered governance also integrates directly with technical reviews. Each governance tier aligns with specific review milestones that assess both Agile progress and systems engineering compliance. For instance, the Stakeholder Needs Review (SNR) validates that Agile objectives align with mission needs; the System Requirements Review (SRR) ensures that requirements decomposition and traceability are complete; and the Preliminary Design Review (PDR) evaluates whether the system architecture supports Agile increments without violating design constraints.

These reviews are synchronization points where teams demonstrate evidence of progress, verification, and alignment and should not be considered project canceling gate events. They act as calibration mechanisms, adjusting the balance between flexibility and control based on current system maturity. When implemented correctly, this approach replaces the adversarial tone of traditional reviews with a collaborative assessment of readiness, risk, and adaptability.

4.7.6 Tiered Governance in Practice

A practical example of tiered governance can be found in large defense software modernization efforts. At the enterprise level, Strategic Governance defines compliance with cybersecurity frameworks such as Risk Management Framework (RMF) and ensures that all programs maintain Authority to Operate (ATO) status. At the program level, Programmatic Governance maintains technical baselines in Model-Based Systems Engineering (MBSE) repositories and synchronizes release schedules across multiple Agile teams. At the team level, Project governance manages Sprint execution, backlog prioritization, and continuous integration pipelines through programs such as GitLab and Jira.

Changes flow through this system seamlessly. A developer proposing a modification to an interface schema submits a change request at the team level. The request triggers an automatic assessment against baseline dependencies maintained at the program level. If no external impacts are detected, the change is executed within the Sprint. If impacts exist, the change escalates to the program board for controlled integration. This tiered decision pipeline preserves agility while ensuring that no change violates traceability or certification integrity.

4.7.7 The Outcome: Controlled Agility

Through tiered governance, SEFAD achieves what traditional methodologies could not: controlled agility. This means that flexibility is no longer synonymous with risk, and compliance is no longer synonymous with stagnation. Every actor in the system operates with a clear understanding of both their freedom to act and their obligation to maintain traceable accountability, whether developer, tester, architect, or program manager. This increases flow of work, minimizing bottlenecks and work stoppages.

In practice, this governance model reduces program friction, accelerates delivery, and enhances trust between engineering teams and oversight authorities. It allows organizations to meet both Agile performance goals and the formal assurance expectations of DoD, NASA, and FAA stakeholders. Tiered governance, therefore, is not simply an administrative mechanism. It is the structural backbone that enables SEFAD to embody its central philosophy: that agility and rigor are not opposites, but complements, when organized by disciplined engineering design. Agile thrives on change, but critical systems require control. SEFAD achieves both through tiered governance.

4.7.8 Configuration Control Board (CCB) Example Structure

All requirement changes, deletions, or additions are reviewed by a multidisciplinary CCB composed of engineering, quality assurance, cybersecurity, and program management representatives.

Each change is evaluated against four dimensions:

- Cost impact
- Schedule impact
- Performance impact
- Risk and safety implications

Change workflow example:

1. Developer identifies need for modification during Sprint.
2. Product Owner drafts a Change Request (CR) linked to affected requirement.
3. CCB evaluates CR in weekly cadence.
4. Approved changes propagate through the digital thread automatically.

This process ensures that Agile adaptability operates within controlled parameters, aligning with DoDI 5000.88 and ISO/IEC/IEEE 15288 standards for configuration management.

4.8 Documentation and Knowledge Retention

Documentation is not an enemy of agility. It is a repository of organizational learning and system intent. In traditional software engineering environments, documentation has often been viewed as static, voluminous, and disconnected from real progress, often existing in disconnected artifacts that are often not kept up to date and poorly traced to one another. SEFAD reframes this perspective entirely by treating documentation as a living product of the engineering process that evolves continuously as the system evolves. Instead of being produced at the end of a project to satisfy auditors, documentation becomes an integral part of every sprint, ensuring that design decisions, rationale, and test results are captured in real time. This approach transforms documentation from a bureaucratic obligation into a tool of transparency, traceability, and technical maturity (Fig. 4.9).

SEFAD integrates documentation into the Agile workflow:

- User stories generate lightweight design notes: Each user story not only defines a functional outcome but also prompts short, focused annotations describing implementation logic, interfaces, and constraints. These micro-documents build the design record iteratively without interrupting Agile flow.
- System models (SysML, Wireframes, DoDAF) evolve concurrently: Instead of static diagrams frozen at baseline, SEFAD treats models as living artifacts that evolve sprint by sprint. They automatically reflect design refinements, requirement updates, and interface adjustments. This concurrent modeling maintains system-level visibility without constraining adaptability.
- Verification data automatically populate compliance reports: Test results, validation metrics, and coverage evidence are captured as part of the development pipeline. Automated scripts feed this data into formatted compliance reports, which reduces manual reporting overhead while maintaining audit readiness.
- Baseline artifacts are versioned through digital configuration tools: Version-controlled repositories such as Git, Cameo Teamwork Cloud, or DOORS Next

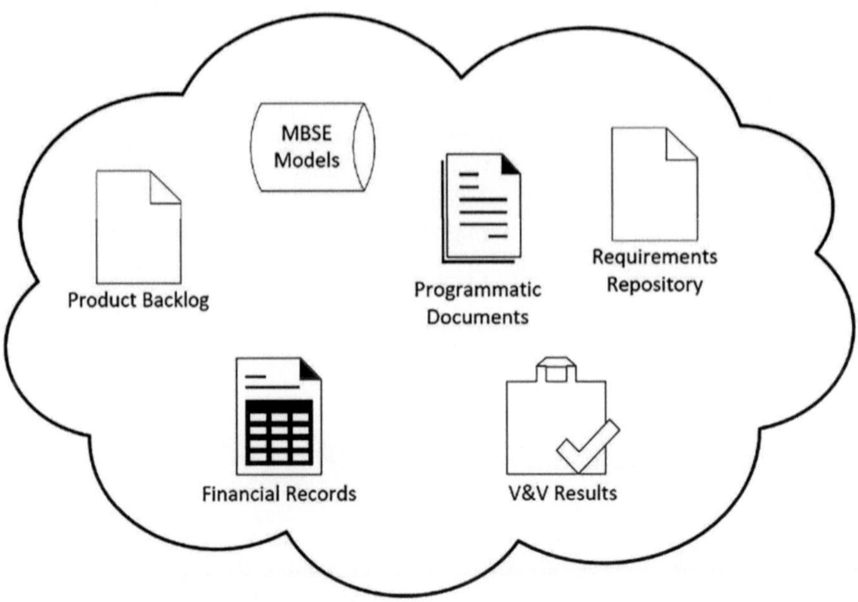

Fig. 4.9 Project documentation

maintain digital baselines of every artifact including requirements, models, test cases, and documents. This ensures that traceability and configuration integrity are preserved even as teams iterate rapidly across distributed environments.

By embedding documentation into every iteration, SEFAD safeguards the institutional memory of the system. Future maintainers, auditors, and system engineers can reconstruct not only what was built but also why it was built that way. Design rationale, trade studies, and test evidence remain linked to the evolving digital thread, enabling confident sustainment, certification, and modernization years after initial deployment. In this model, documentation becomes a living bridge between agility and accountability. It ensures that the pursuit of speed in development does not come at the cost of transparency or long-term system understanding. This is especially important in critical systems in the defense and financial industries, which might stay in use for decades. In these systems, the original engineers and project managers might have moved on from the project. Therefore, having good documentation on requirements, rationale, and answering the "whys" and "hows" of the system are incredibly important.

4.9 Challenges and Mitigations

4.9.1 Requirements Volatility

Agile's ability to adapt quickly to evolving customer needs is one of its greatest strengths in commercial development. However, in the context of critical systems, that same flexibility can introduce significant risk. Frequent requirement changes can destabilize baselines, disrupt test plans, and create inconsistencies between hardware, software, and mission documentation. In systems where safety, reliability, and regulatory compliance are paramount, volatility in requirements can cascade into schedule slips, cost overruns, or certification failures. What is beneficial in a dynamic product environment becomes a liability in a mission-critical one if not properly controlled. It is therefore of utmost importance to address this volatility for critical system development.

SEFAD addresses this challenge through a disciplined requirements management approach that distinguishes between final and conditional requirements. Final requirements are stable, fully vetted, and formally aligned to a specific release baseline. They represent system capabilities that must not change without formal impact assessment and configuration control. Conditional requirements, by contrast, capture evolving or uncertain needs that may benefit from iterative discovery. These are permitted to change within controlled review cycles, allowing Agile teams to experiment and refine without jeopardizing the integrity of the release baseline. In fact, controlled requirements can be completely deleted without affecting a release. This allows customers the flexibility to communicate their requirements knowing that it is okay to pivot the development path further in the development schedule.

One way to visualize this idea is through the software development repository management process of "dirty trunk, clean branch." In this method, one imagines a tree laying on its side. The trunk of said tree is the core repository for code. This is our "dirty trunk" in that the developers will be checking code in and out of this repository. To maintain version control, at releases, developers copy specified code that is applicable to a specific release into its own repository. This becomes a branch on the tree, representing the code base at the time of a release. Applying the same principles to a requirements repository is then simple. The main repository acts as the trunk and holds all requirements that are to be worked on, specifically, it includes all the requirements, including the conditional requirements. Upon release of an iteration, the requirements aligned with that release become final and are then saved to a specific baseline, or branch, repository. Branches never include conditional requirements. This allows the development team to keep a complete repository of final and conditional requirements while maintaining configuration and version control of the system. An example of this method is shown in Fig. 4.10.

This two-tiered classification allows SEFAD to preserve adaptability while maintaining engineering rigor. It enables organizations to respond to stakeholder input and technical discovery in a structured way, ensuring that flexibility does not erode traceability or verification confidence. By embedding this categorization within each Agile iteration, SEFAD ensures that change is deliberate, traceable, and bounded by mission assurance principles. In this way, the model converts Agile's natural volatility into a managed process that supports both development flexibility and system stability.

4.9.2 *Tooling Integration*

Effective SEFAD execution depends on a fully integrated digital tool chain that unifies the traditionally separate domains of requirements management, Agile development, and model-based systems engineering. In many organizations, these tools operate in isolation, creating manual handoffs, duplicate data entry, and opportunities for traceability loss. SEFAD eliminates these gaps by establishing direct interoperability among systems such as requirements repositories (e.g., IBM DOORS Next), Agile management tools (such as JIRA or Azure DevOps), and modeling environments (such as Cameo Systems Modeler or Enterprise Architect).

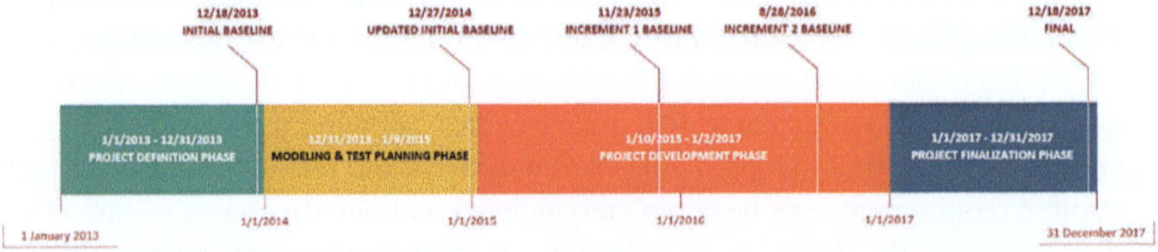

Fig. 4.10 Dirty trunk requirements baseline example

Through this integration, every user story, requirement, and design model becomes part of a synchronized digital ecosystem where updates propagate automatically across tools.

This connected model-based architecture is essential because it enables real-time traceability from stakeholder needs to system design, implementation, and verification evidence. When a requirement changes, linked user stories, model elements, and test cases update accordingly, preserving configuration integrity without requiring labor-intensive manual reconciliation. Similarly, Sprint-level activities in Agile management tools can feed progress metrics directly into programmatic reports, ensuring that technical advancement remains aligned with contractual and regulatory obligations. The integration also supports automated compliance generation, allowing documentation packages for reviews or audits to be produced directly from live data rather than recreated by hand.

Automation of traceability and compliance evidence is not merely a convenience; it is the mechanism that preserves agility under rigor. Without it, Agile teams in critical systems environments become burdened by documentation overhead and lose the speed which makes iterative development effective. Automated traceability links, embedded tests, and dynamically generated compliance matrices allow teams to maintain both responsiveness and accountability. In this way, SEFAD transforms the digital tool chain from a set of disconnected utilities into a coherent assurance framework, one that continuously proves alignment between stakeholder intent, system behavior, and verified performance.

4.10 The Future of Agile Requirements Management

As defense and aerospace systems evolve toward model-based and digital engineering, requirements will increasingly exist as dynamic model elements rather than static text. The core idea of a requirement being a statement in clear, concise, and plain language is changing. Requirements come in many forms now, not least of which are model-based diagrams and simulations. SEFAD anticipates this shift, embedding requirements directly into system models, linking "shall" statements to executable architectures, simulations, and verification scripts. The ultimate goal is not documentation, but decision integrity. We will cover more of this in Chap. 5.

4.11 Conclusions

Agile Requirements Management within SEFAD represents a deliberate fusion of two traditionally opposing philosophies: the adaptability of Agile and the discipline of systems engineering. Agile methods thrive on flexibility and iterative feedback, allowing teams to respond quickly to emerging needs and stakeholder insights. Systems engineering, in contrast, emphasizes control, traceability, and formal verification to ensure that every system function fulfills its intended purpose under defined constraints. SEFAD integrates these paradigms into a single, cohesive approach that preserves the strengths of both. It enables teams to move fast without losing the structure necessary to guarantee safety, performance, and compliance in critical systems.

At the heart of this integration is SEFAD's principle of requirement grounding. Every user story is explicitly tied to a verifiable "shall" statement that captures the formal system requirement in precise, testable language. This linkage bridges the traditional gap between Agile's informal narratives and the rigor of specification-driven engineering. In doing so, SEFAD ensures that responsiveness to change never comes at the expense of accountability. Each user story becomes both an operational objective and a traceable requirement artifact, fully aligned with higher-level system intent and validation criteria.

SEFAD further strengthens this alignment by incorporating INCOSE's attributes of good requirements into the Agile workflow. Each statement is evaluated to confirm that it is necessary, verifiable, feasible, and traceable to stakeholder needs. These attributes serve as continuous quality gates within iteration planning and review, transforming the backlog from a dynamic wish list into a structured, auditable repository of system intent. As a result, requirement quality is not deferred to late-stage documentation; it is validated continuously as part of the sprint cycle. This discipline creates measurable assurance that the system being built remains aligned with both user expectations and mission-critical constraints.

Agility without rigor breeds chaos, leading to rework, technical debt, and loss of confidence. Rigor without agility breeds stagnation, creating bureaucratic inertia that resists adaptation in the face of new information. SEFAD reveals that true engineering excellence lies between these extremes. It defines a sustainable equilibrium where creativity and discipline coexist, allowing organizations to innovate quickly but never blindly. In this balanced environment, change is embraced but always verified, and progress is accelerated without ever compromising the integrity of the system or the trust of its stakeholders.

4.12 Discussion Questions

Balancing rigor and agility: How does SEFAD's fusion of predictive discipline and Agile flexibility challenge the traditional assumption that rigor and adaptability are mutually exclusive? What trade-offs arise when attempting to sustain both simultaneously in a defense or aerospace program?

Requirements as architectural DNA: SEFAD defines requirements as the "architectural DNA" of a system. In practice, how can program teams ensure that evolving user stories and model-based artifacts remain genetically consistent with these core requirements as the system matures?

Applying the INCOSE attributes continuously: Traditional reviews (SRR, PDR, and CDR) evaluate requirements at fixed milestones. SEFAD applies the INCOSE ten attributes to all requirements during every iteration. What organizational or cultural changes are required for engineering teams to treat these attributes as living evaluation criteria rather than checklist compliance items?

Traceability and the digital thread: Discuss how digital thread integration reshapes accountability and audit readiness from stakeholder need through verification evidence. How might automated traceability influence the role of the systems engineer in an Agile environment?

The "shall" statement vs. the user story: In SEFAD, every user story must trace to a "shall" statement. What risks arise if this linkage weakens or breaks, and how can tooling or process design prevent such traceability loss without stifling team autonomy?

Controlled agility through tiered governance: Tiered governance distributes authority across strategic, programmatic, and project levels. How does this model transform governance from a control mechanism into an enabler of flow? Provide examples of where decision latency can be reduced without sacrificing safety or compliance.

The dirty trunk/clean branch analogy: Evaluate the "dirty trunk, clean branch" analogy for managing final and conditional requirements. What are the strengths and weaknesses of applying version control metaphors from software engineering to requirements repositories?

Continuous verification and validation: SEFAD embeds V&V activities in every sprint rather than reserving them for end-of-cycle testing. How does this approach change defect discovery patterns, stakeholder engagement, and certification readiness compared to the classical V-model?

Documentation as a living artifact: SEFAD reframes documentation as a living product of the engineering process rather than a static deliverable. What implications does this have for knowledge retention, workforce turnover, and long-term sustainment of critical systems that may outlive their original teams?

Future of requirements management: As model-based and digital engineering practices evolve, textual "shall" statements are increasingly embedded within executable architectures. How might this convergence of models and requirements redefine the very concept of a requirement, and what new verification methods will be necessary to maintain decision integrity?

Reference

1. INCOSE Requirements Working Group, *Guide to Writing Requirements* (International Council on Systems Engineering (INCOSE), San Diego, 2023)

Chapter 5
The Use of Model-Based Systems Engineering in Agile Software Development

5.1 Introduction

As modern software systems become more complex and interconnected, the limitations of traditional document-based engineering approaches have become increasingly clear. Manual methods of requirements documentation, design and architecture representation, and verification and validation efforts often result in inconsistencies, communication gaps, and delayed discovery of defects. This is often due to the nature of these distinct, non-connected artifacts becoming rapidly desynchronized. When each project member or team keeps their own copy of an artifact, artifacts do not maintain synchronicity. These issues are amplified in systems that must evolve quickly to meet emerging mission demands or security threats. Model-Based Systems Engineering, commonly known as MBSE, provides a structured solution by replacing static and desynchronized documentation with interconnected, digital system models. Through this model-centric approach, engineers gain the ability to visualize relationships, dependencies, and constraints in real time, allowing for faster and more accurate decision-making across development teams.

Within Agile software development environments, MBSE plays a crucial role in maintaining discipline and traceability without hindering flexibility. Agile methods emphasize iterative progress and customer collaboration, yet they can struggle to ensure that every feature increment aligns with overarching system goals. MBSE addresses this challenge by serving as a unifying framework that integrates system architecture, functional behavior, and verification and validation logic into a single, coherent, synchronized digital engineering ecosystem. In doing so, it preserves the adaptability of Agile practices while introducing the rigor necessary for high-assurance systems. The result is a development environment where iterative changes are guided by data-driven models that accurately reflect stakeholder intent and system constraints.

D. Rosson, *Systems Engineering in Agile Software Development*, https://doi.org/10.1007/978-3-032-14613-7_5

Wheaton and Herber (2024) describe digital requirements engineering as a natural evolution of MBSE that establishes an Authoritative Source of Truth, often referred to as an ASoT, for all system artifacts [1]. This unified source ensures that every requirement, design element, and verification activity remains synchronized throughout the lifecycle. When applied within Agile workflows, the digital thread formed by MBSE connects evolving requirements to ongoing validation and testing activities. Teams can trace the impact of design decisions immediately, reducing rework and improving confidence in system integrity. In critical software systems, this model-based integration ensures that each user story or product increment is validated not only for functionality but also for compliance with mission, safety, and performance objectives.

Transitioning from the limitations of traditional document-based methods to a digitally integrated engineering environment sets the stage for a broader transformation in how systems are conceived, developed, and sustained. As organizations increasingly adopt Agile methodologies to meet the demands of rapid iteration and evolving stakeholder needs, the question becomes how to preserve the discipline and traceability essential for complex, safety-critical systems. This is where the role of MBSE becomes pivotal. MBSE is not merely a technical enhancement but a cultural and procedural bridge which links the structure and rigor of systems engineering with the flexibility and speed of Agile delivery. Understanding how MBSE supports, informs, and accelerates Agile transformation is therefore essential to realizing a true digital engineering ecosystem that balances adaptability with assurance.

5.2 The Role of MBSE in Agile Transformation

MBSE extends Agile's foundational principles of collaboration, rapid feedback, and continuous adaptability into a model-driven environment where each iteration is supported by a living system model. Rather than displacing Agile methods, MBSE augments them by providing the structure and traceability that Agile alone often lacks in safety or mission-critical contexts. It is through MBSE that a full understanding of the system and problem domain emerges, lowering risks for the development team, project managers, and customers. Communication becomes easier as knowledge is translated into diagrams, documenting the current understanding of requirements to architecture and then onto functions, allowing developers to easily display onscreen what they want to accomplish and how they will do so. Customers generally have a much easier time understanding the proposed system architecture when shown graphically, allowing for rapid feedback. While there is an upfront cost in time and money for the development of MBSE architectures, it is usually recouped later in the development cycle as less rework is required.

Herber and Eftekhari-Shahroudi demonstrated this model-driven synergy through the implementation of Model-Based Structured Requirements (MBSR), a method that integrates SysML representations directly into Agile workflows [2].

This approach redefines how requirements are created, traced, and validated within iterative development environments. Instead of static textual statements that quickly fall out of sync, each requirement is expressed as a structured SysML element, embedding behavioral, structural, and parametric information within the same authoritative model. These elements become both design artifacts and verification anchors, enabling engineers to trace functionality, constraints, and test criteria through the full system lifecycle. By representing requirements in a shared, executable model, MBSR ensures that interdisciplinary teams are referencing the same synchronized source of truth. This integration minimizes ambiguity, accelerates change impact analysis, and allows for automated validation and consistency checks throughout development. Importantly, it aligns with Agile principles by keeping feedback loops short: when a user story evolves, the corresponding SysML requirement updates immediately, ensuring architectural alignment without delaying iteration. In practice, this allows Agile development to remain flexible while ensuring that every iteration builds upon validated architectural foundations.

The U.S. Department of Defense (DoD) and NASA have both formally recognized MBSE as a cornerstone of their ongoing digital engineering transformation efforts. Within their respective digital acquisition and modernization strategies, MBSE is positioned not merely as a documentation improvement tool but as the foundational enabler of Agile, data-driven system development. The DoD's *Digital Engineering Strategy* emphasizes the creation and sustainment of an Authoritative Source of Truth (ASoT), which is defined as "a living system model that supports iterative development cycles and allows teams to evaluate design changes continuously, rather than relying on static specifications or late-stage integration testing" [3]. Similarly, NASA's *Digital Transformation Initiative Strategic Framework & Implementation Approach* integrates MBSE into its Agile project management frameworks, where system models are used to assess technical risk, verify design intent, and synchronize multidisciplinary engineering teams across concurrent development efforts [4]. By embedding MBSE into these Agile-aligned strategies, both agencies seek to reduce rework, shorten feedback loops, and improve traceability from stakeholder intent through verification. Unified models replace fragmented documentation, ensuring that every design iteration and sprint increment is informed by verified, current system data rather than untested assumptions or legacy artifacts. This digital-first approach transforms the role of systems engineering within Agile environments from a gatekeeping function at phase transitions to a continuous validation process that supports rapid adaptation without compromising rigor. In doing so, MBSE becomes not just a technical framework but a cultural bridge, aligning government engineering practices with the collaborative, iterative, and value-driven principles that define modern Agile transformation. Agile software teams operating in such domains thus rely on MBSE as the formal backbone of continuous engineering assurance.

5.3 SysML as the Foundation for Agile Modeling

The Systems Modeling Language (SysML) remains the industry standard for system definition, analysis, and communication across complex engineering domains. Built upon the Unified Modeling Language (UML), SysML extends its capabilities to represent both structural and behavioral aspects of a system within a unified, model-based framework. This includes defining system requirements, functional decompositions, physical architectures, state behaviors, and parametric relationships within interconnected diagrams that form a holistic representation of the system's intent. Through the integration of these capabilities presented in diagrams, SysML enables engineers to visualize not only what a system is but also how it behaves and how its components interact under varying conditions.

Within an Agile development context, SysML becomes the vital bridge between high-level system intent and the low-level software and hardware implementations that deliver functionality. For software teams, the high-level intent is traditionally mapped in epics, features, and user stories, often resulting in non-documentation of the traceability between user stories and implementation. While Agile emphasizes iterative delivery and rapid feedback, SysML ensures that every iteration remains anchored in the broader system architecture, preventing drift from the original mission and technical objectives. By mapping user stories directly requirements that are then traced to SysML model elements, Agile teams can maintain continuous alignment between evolving stakeholder needs and verified system designs.

Furthermore, SysML's capability to represent relationships through traceability links supports the Agile principle of transparency and adaptability. When requirements change, engineers can immediately visualize the downstream effects on interfaces, subsystems, and test cases, thereby allowing informed decision-making during sprint planning and backlog refinement. This model-driven traceability ensures that agility does not come at the expense of system integrity, fostering an environment where incremental development, continuous integration, and rigorous systems thinking coexist.

As noted by Wheaton and Herber in "Digital requirements engineering with an INCOSE-derived SysML meta-model," SysML's extensibility allows it to capture "structured requirements" directly linked to the system model, replacing ambiguous natural language requirements with formal, model-referenced statements [1]. The use of SysML stereotypes and meta-model extensions enables requirements engineers to express obligations such as "The control module shall transition to safe mode within 2 s under enclosure door open conditions" as verifiable, connected elements within the system model. Wheaton and Herber's ISO and INCOSE-derived SysML meta-model further refines this approach, ensuring that each "shall" statement follows standardized syntactic patterns that align with the INCOSE Guide to Writing Requirements and ISO/IEC/IEEE 29148:2018. This structured syntax not only improves consistency but also facilitates automated validation and verification, supporting Agile teams in maintaining continuous compliance.

5.4 Model-Based Structured Requirements (MBSR) in Agile Contexts

MBSR represents one of the most significant advances in the ongoing integration of MBSE and Agile methodologies. Traditional requirements, typically stored as static text within documents or databases, often suffer from inconsistency and lack of synchronization with evolving system models as individual documents get updated individually but not as a whole. For example, if a requirement statement is updated and not the associated drawing, then the requirements become out of sync. MBSR directly addresses these challenges by providing a structured, model-centric approach that links textual requirements to SysML-based representations of system behavior, structure, and constraints. This approach preserves the human readability necessary for stakeholder communication while enabling the machine-verifiable precision required for automated validation, consistency checking, and digital traceability. Through MBSR, every requirement becomes a living element within the system model, no longer existing as an isolated statement but as an integrated component tied to specific design artifacts, test cases, and verification activities. This dual representation bridges the gap between natural language intent and executable system logic, ensuring that the semantic meaning of each requirement is captured both descriptively and formally. As a result, MBSR facilitates automated reasoning, impact analysis, and bidirectional traceability, empowering Agile teams to respond to change without sacrificing alignment with mission objectives or regulatory standards. Within Agile environments, MBSR provides a practical mechanism for connecting user stories and backlog items to the underlying model architecture. This linkage allows teams to iterate rapidly while maintaining systemic coherence as developers can adapt Sprint work items to new stakeholder inputs while remaining anchored in validated model structures. By embedding requirements directly into the digital engineering ecosystem, MBSR transforms traditional documentation into an active, authoritative source of truth, enabling continuous verification throughout the development lifecycle and embodying the Agile principle of delivering value through frequent, model-informed iteration, ensuring that adaptability and rigor reinforce rather than oppose one another.

In MBSR, each structured requirement follows a pattern. This pattern captures actors, functions, performance measures, and operating contexts in a single, cohesive construct. The statements are worded in the following manner: [Who] shall [What] [How Well] under [Condition]. An example SysML Requirement Block that meets the MBSR structure is shown in Fig. 5.1. Note that in this case, the context of "under" has been replaced by "between," which serves the same purpose of specified elaboration in this statement. Baseline requirements are traced to custom "structured requirements." These requirements include traceability to the Who, What, How Good, and Condition, in the form of system artifacts as well as linkages through the original requirements to user stories. This allows for the original requirement to be untouched and all the functionality inherent in SysML to be maintained

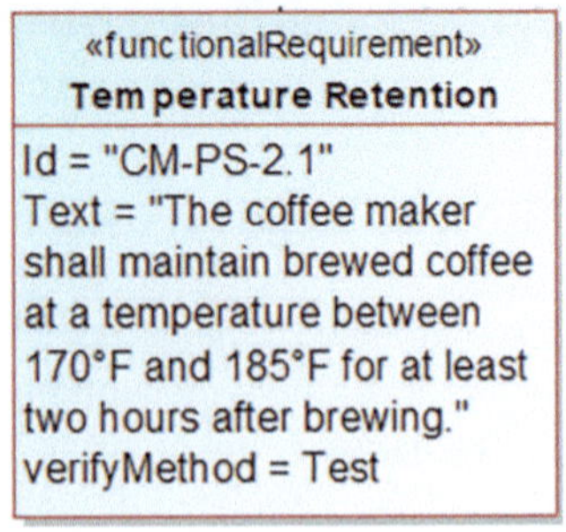

Fig. 5.1 MBSR statement example

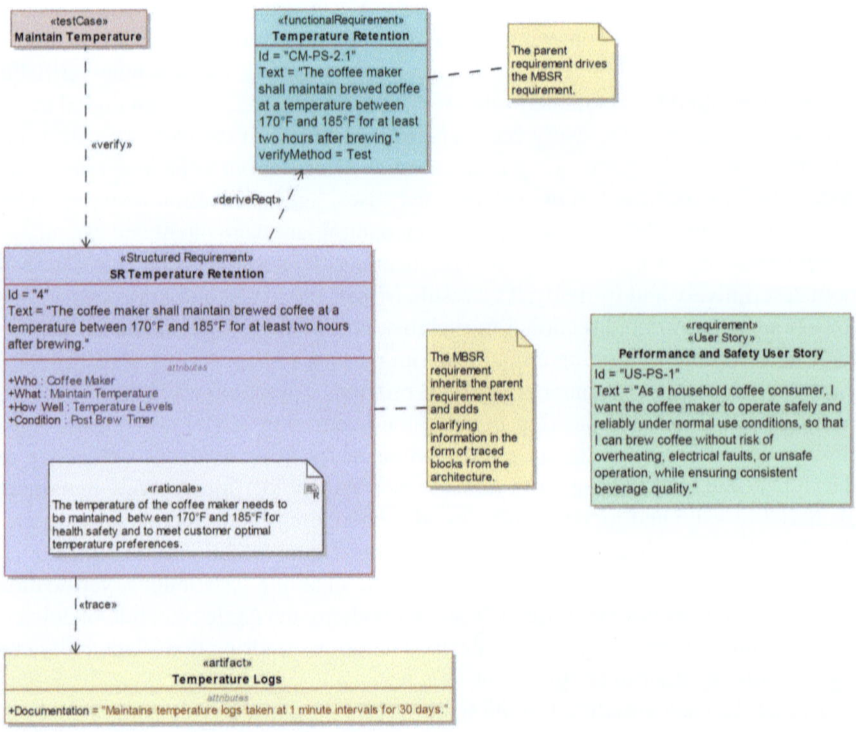

Fig. 5.2 MBSR requirement modeling structure

while adding a layer of traceability that is not traditionally available. An example of this can be seen in Fig. 5.2.

In Agile software teams, these structured requirements can augment user stories by providing direct traceability from user intent to verifiable model behavior. For instance, a traditional Agile story such as "As an operator, I need the system to alert me of power fluctuations" can be translated into a SysML-linked structured requirement that specifies response time, alert thresholds, and performance boundaries, all of which can be traced and automatically connected to testing and simulation models. While this increases time spent developing the model, over the lifecycle of a system, it will decrease the time required to perform testing and increase the rigor of the associated verification and validation activities.

5.5 Model-Based Software Systems Architecture

Model-Based Systems Engineering (MBSE) provides a structured, model-centric approach to documenting and designing critical software systems with precision and traceability. SysML's diagrammatic constructs, Block Definition Diagrams (BDDs), Internal Block Diagrams (IBDs), Activity Diagrams, and Sequence Diagrams, offer a powerful means to capture both the architecture and behavior of software systems in a form that increases both engineering and technical rigor. When applied to software development documentation, these diagrams become the backbone of a digital engineering environment, replacing static design documents with dynamic, interconnected artifacts. As a centralized repository of knowledge, updates are instantly promulgated out to all model users, removing the barrier of communication of static and physical documentation. This model-based documentation ensures that architecture, logic, and interfaces remain synchronized throughout the lifecycle, supporting Agile iteration, design verification, and sustainment activities with unparalleled fidelity.

Block definition diagrams establish the high-level structure of the software system and are often used in place of traditional software architecture descriptions. Within software documentation, the BDD becomes the Software Architecture Overview section by defining major software components, data objects, and their hierarchical relationships. For example, a BDD can define subsystems such as the data management layer, communication layer, and user interface layer, each represented as blocks with standardized interfaces. These diagrams can be integrated into system design documentation to provide a clear structural taxonomy and serve as the authoritative source for software modularization, component ownership, and integration planning. Additionally, these levels can be abstracted, allowing for multiple views and viewpoints of system components and levels of technical depth. This can be further expanded to various system description diagrams, allowing for version differences and configuration controls.

An example BDD for a coffee machine is shown in Fig. 5.3. This example shows how multiple layers of systems and subsystems can be displayed in one diagram. For embedded systems, this can easily show the relationships between hardware and software systems. For software systems, not only can the system structure be shown, but also the relationships to users, hosting infrastructure, and communication paths. This creates easy to understand documentation for the system that is often spread out over multiple disconnected documents, creating an ease of knowledge transfer between software developer, database administrator, and IT professional.

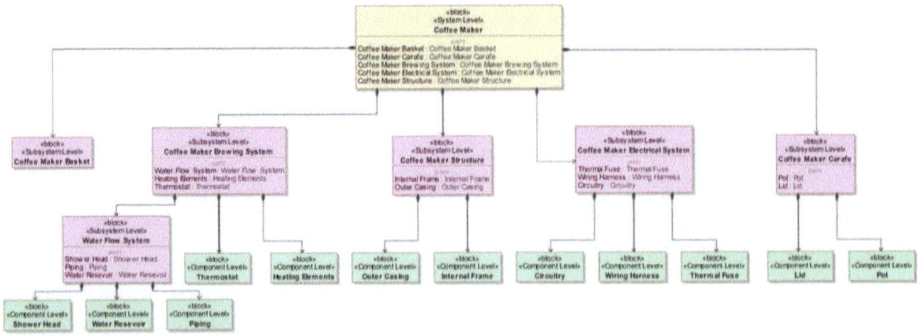

Fig. 5.3 Coffee maker SysML system diagram example

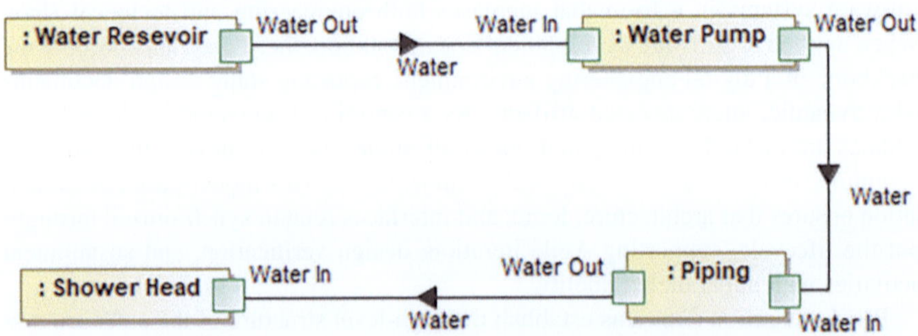

Fig. 5.4 Coffee maker water flow system internal block diagram

Internal block diagrams expand upon this foundation by capturing the details of how software components interact at runtime, effectively representing the *Interface Control Documentation* (ICD) in a visual, model-based form. IBDs specify ports, connectors, and data exchanges between modules, clarifying communication protocols, data types, and control signals. IBDs easily display flows of data, material, signals, and other information between components in a system. This can be seen in Fig. 5.4, where the flow of water from the system water reservoir out through the shower head. When embedded within development documentation, they replace static interface tables with live, version-controlled diagrams that can be auto-generated from the system model. This practice enhances maintainability from which engineers can trace interface definitions to source code or middleware components, ensuring consistency between design intent and implementation. In critical systems, IBDs serve as a compliance artifact for interface verification and software integration readiness reviews, often meeting required documentation requirements without additional effort.

Activity Diagrams translate directly into behavioral documentation traditionally found in *Software Design Descriptions (SDDs)* or *Functional Flow Block Diagrams (FFBDs)*. They describe the logical flow of operations, decision branches, and parallel processes within the software, making them ideal for documenting control logic, algorithms, and error handling routines. Within software documentation, activity diagrams provide developers and testers with a visual blueprint of operational logic, enabling alignment between design and test case development. Moreover, activity diagrams can be linked to simulation tools or model-based testing environments, creating executable documentation where workflows can be verified dynamically rather than interpreted statically.

Figure 5.5 shows how an activity is displayed in SysML. When a coffee user wants to brew coffee, they follow a procedure. A simplified form of this is to turn on the coffee machine and select a brew cycle. The machine then will brew the coffee until a level has been met. Once that is accomplished, the machine maintains the coffee temperature at a specified level. Each of the rounded boxes is an "activity." Each of these activities can have sub-activities, drilling down as far as the systems engineer desires to model. Through the use of these diagrams, behaviors can be

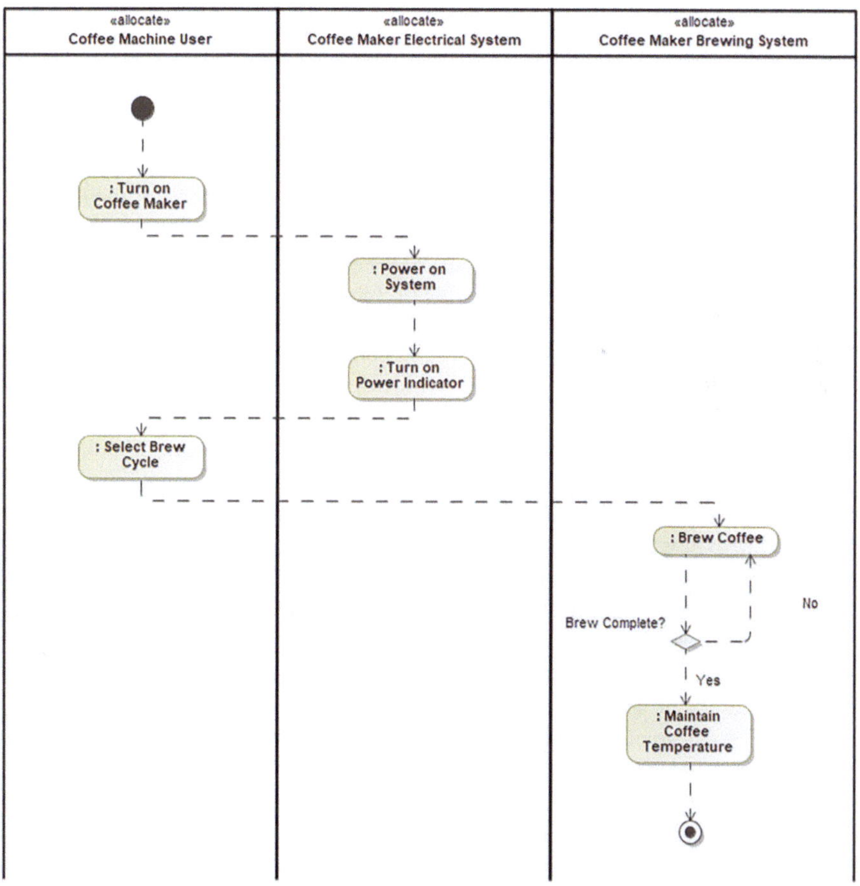

Fig. 5.5 Brew coffee activity diagram

modeled in a similar fashion to what many developers commonly call "scenarios," as they both describe system behavior from the perspective of flow and interaction. The emphasis on these diagrams is on what happens, who or what performs it, and in what order. The biggest difference, and the true power of activity diagrams over scenarios, is that scenarios are written in natural language whereas activity diagrams are formal, unambiguous visual models. As has been stated many times, humans are very visually oriented and can take in a large amount of information quickly. Seeing activities in this structure makes understanding system behaviors much easier for most engineers and stakeholders

Sequence Diagrams capture the temporal dimension of software behavior and are often used to document *Software Interaction Scenarios* or *Use Case Realizations*. They depict message exchanges between objects or subsystems over time, illustrating timing, ordering, and causality. Within formal documentation, these diagrams clarify how software components coordinate during operations such as initialization, fault recovery, or communication with external systems. Because sequence diagrams explicitly model synchronization and latency, they support verification of real-time constraints and concurrency handling which are crucial for safety-critical and mission-critical applications. When linked to model repositories, sequence diagrams can also be auto synchronized with simulation results or test logs, ensuring the documentation reflects actual system performance.

The sequence diagram shown in Fig. 5.6 displays the steps in communication required by the components of the Coffee Maker Brewing System to maintain the temperature of the brewed coffee. After the brew cycle has completed, the Coffee Maker Brewing System sends a message signal to the Thermostat to begin the process of maintaining the coffee temperature. This is done by calling the MaintainTemperature() function in the software of the Thermostat component. The MaintainTemperature() function performs a constant loop, continuously checking the temperature of the coffee. If the temperature falls below 170 °F, then the Thermostat sends a signal to the Heating Elements to turn on. The Heating Elements send a signal when heating has started. The Thermostat continues monitoring the temperature. If it then rises above 185 °F, the Thermostat sends a signal to stop heating to the Heating Elements. Again, when the elements turn off, the component sends a response message that it has done so. This continues until the MaintainTemperature() function changes to "false." When that occurs, the Thermostat sends a signal to turn off the Heating Elements, which then responds. Finally, the Thermostat sends a message to the Coffee Maker Brewing System informing the subsystem that the MaintainTemperature() function is complete. Through the use of sequence diagrams in this manner, software class interactions can easily be documented and understood, as well as system and component interactions in embedded systems.

Together, these SysML diagrams create a living ecosystem of software documentation. Instead of static design documents that quickly fall out of date, engineers maintain an integrated and centralized model representing up-to-date architecture diagrams, interface definitions, and behavioral logic flows. This approach embeds documentation directly within the development workflow, ensuring that every

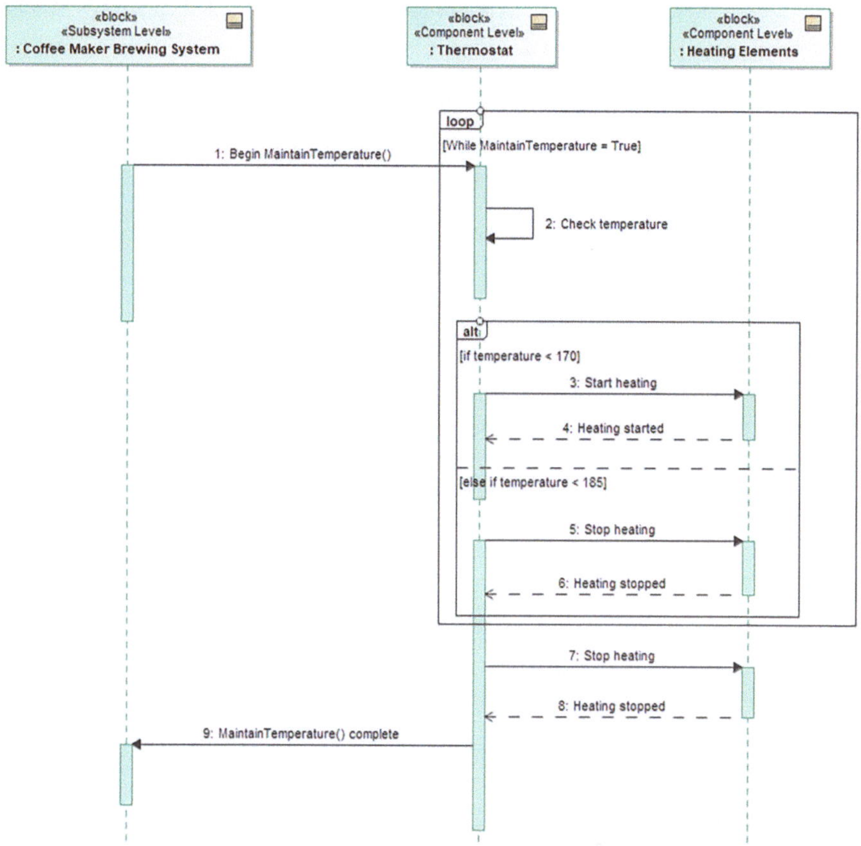

Fig. 5.6 Coffee maker maintain temperature sequence diagram

change to the software model propagates to all dependent views. For critical software systems, this model-based documentation paradigm not only enhances engineering and technical rigor but rather institutionalizes it, creating an auditable, analyzable, and continuously evolving record of system design integrity.

5.6 The Digital Thread and Continuous Verification

One of the transformative benefits of combining MBSE with Agile is the establishment of a requirements digital thread. This can be thought of as a continuous, traceable chain connecting stakeholder needs, system models, software code, and verification data. Herber and Eftekhari-Shahroudi (2023) describe this digital thread as a "single-source-of-truth model" that spans the entire lifecycle, from concept through testing [2]. Wheaton and Herber further highlight that digital requirements engineering, when integrated with MBSE, enables automated traceability checks and model-based verification [1]. This ensures that every system change preserves its relationship to validated architecture elements, satisfying INCOSE's expectations for bidirectional traceability. In an Agile environment, this means that each sprint's deliverables are verified against defined requirements through the results of unit tests and traditional testing. Verification results, model changes, and code artifacts feed into a common digital ecosystem, allowing for near-real-time impact analysis. This integration supports the Agile principles of rapid feedback and transparency while also addressing systems engineering's demand for continuous assurance.

In software, digital threads are not often discussed but are very important. Figure 5.7 shows a visualization of a common digital thread for software engineering. Requirements form the foundation and beginning point of the digital thread. Requirements directly drive Verification & Validation, Architecture & Design, and Deployment & Sustainment. The Architecture & Design drive the Infrastructure and Deployment & Sustainment. Infrastructure passes back into Verification &

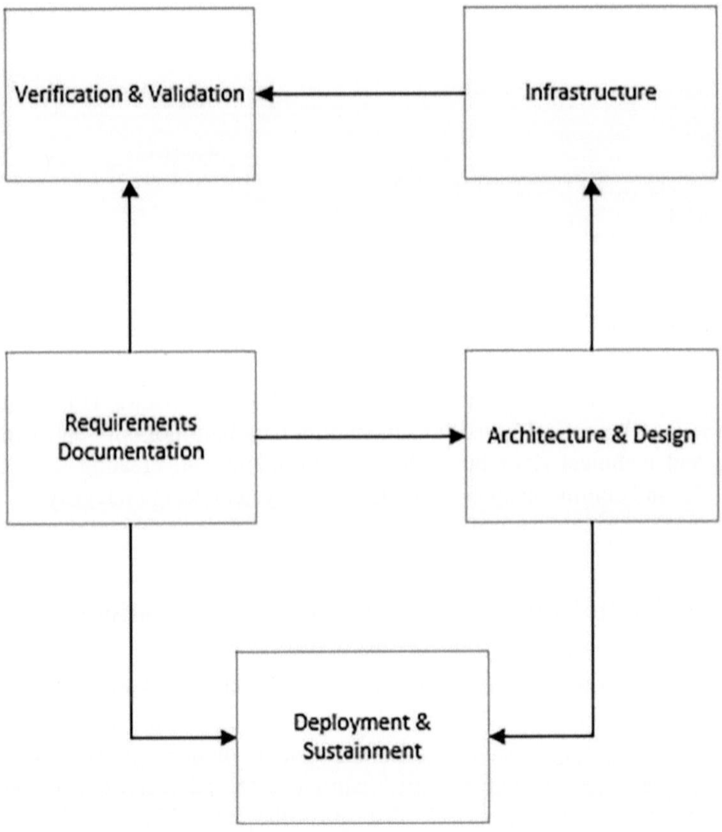

Fig. 5.7 Software engineering digital thread

Validation. The interconnected nature of all of these functional areas is readily apparent when visualized as a digital thread. These pathways can also change dependent upon the development methodologies used.

In software engineering, the digital thread is the integrated data and model-driven framework that connects all digital artifacts, tools, and processes across the software system's lifecycle. It provides a continuous, traceable flow of authoritative information that links system models, software architecture, code, test results, and operational data into a single, coherent engineering ecosystem. In essence, the digital thread ensures that every decision, requirement, and code change is contextually linked to its upstream intent and downstream effects. For software systems utilizing MBSE and digital engineering environments, it acts as the connective tissue between models, data, and decisions, enabling engineers to maintain consistency, traceability, and analytical insight throughout the lifecycle.

5.7 Challenges in Integration

Despite transformative potential of integrating MBSE into Agile software development, the method introduces a fundamental tension between two distinct engineering mindsets. Agile methods emphasize speed, flexibility, and incremental delivery, encouraging teams to prioritize working software and customer feedback over formal documentation. In contrast, MBSE emphasizes discipline, precision, and traceability, seeking to define and govern system behavior through structured, interconnected models. When these approaches collide, teams can struggle to balance the agility needed for rapid iteration with the rigor necessary to maintain model integrity. Therefore, the challenge lies in aligning their principles so that modeling becomes a natural extension of Agile delivery rather than an obstacle to it.

Cultural alignment is only one dimension of this challenge. Successful integration requires the adoption of lightweight modeling frameworks that embed seamlessly into Agile workflows without overwhelming developers. Traditional MBSE practices often rely on more heavyweight processes and formal reviews, which can slow sprint progress. Instead, Agile-oriented modeling must focus on creating just enough architecture by creating model artifacts that are valuable, reusable, and directly linked to code or test cases at a level that brings the highest value. The introduction of automated synchronization tools, such as model-to-code generation, model-driven testing, or bidirectional traceability plugins, helps maintain real-time consistency between design intent and implemented functionality. Equally important is the definition of modeling roles within Agile teams, ensuring that systems engineers, software developers, and testers each understand how modeling supports their responsibilities and sprint objectives.

Beyond process integration, technical interoperability remains a persistent barrier to fully realizing the benefits of MBSE within Agile environments. As Wheaton and Herber observed in NASA applications, even when SysML models adhere to standardized INCOSE-derived profiles, inconsistencies can emerge between modeling tools and requirements repositories such as Cameo Systems Modeler and IBM DOORS Next [1]. These discrepancies often arise from differences in data schema, update frequency, or semantic interpretation across tools. When synchronization errors occur, they can undermine confidence in the "authoritative source of truth," creating confusion about which artifacts, whether it be models, requirements, or code, represent the most current system state. Overcoming this obstacle demands robust data governance, version control, and the use of open interoperability standards like OSLC (Open Services for Lifecycle Collaboration).

Ultimately, integrating MBSE into Agile software engineering is not a one-time transformation but a continuous collaboration across disciplines. Systems engineers must learn to model at the pace of Agile development, while software teams must appreciate the value of models as living documentation that enhances system understanding, traceability, and verification. Product owners play a key role in prioritizing model updates alongside functional increments, ensuring that the digital thread remains intact throughout development. By fostering shared ownership of models and embedding them into daily workflows, organizations can achieve the dual goals of Agile adaptability and MBSE rigor, establishing a unified engineering ecosystem that delivers both speed and systems integrity.

5.8 The Future of MBSE in Agile Systems Engineering

The convergence of MBSE and Agile software development represents a defining evolution in digital engineering that unites speed, precision, and collaboration across disciplines. Traditional software engineering approaches have long separated system modeling from software implementation, creating gaps between conceptual design and operational reality. The fusion of MBSE and Agile closes this divide by embedding system models directly into iterative development cycles, ensuring that architecture, design intent, and verification remain synchronized with working code. This integration marks a shift from static documentation toward living, executable models that evolve as the system does. In doing so, it enables teams to maintain engineering rigor without sacrificing the responsiveness that Agile demands.

Emerging practices such as Agile Model-Based Engineering (Agile MBE) and Continuous Model Integration (CMI) extend DevSecOps principles beyond software code into the modeling domain. Just as continuous integration and delivery pipelines automate testing and deployment, CMI pipelines automate model updates, validation checks, and synchronization with software artifacts. This creates version-controlled, collaborative modeling environments where system models evolve in lockstep with the software they describe. The result is a unified engineering thread connecting requirements, models, and code through a shared digital backbone. In practice, this means model repositories now function much like software repositories, complete with branching, merging, and automated quality checks, allowing engineers to experiment, iterate, and validate architectures with the same agility as code developers.

Academic research, such as that published by Eftekhari-Shahroudi, Herber, and Wheaton, provides an operational foundation for this convergence by demonstrating how INCOSE frameworks for structured requirements can be applied directly within SysML-based Agile environments. Their work shows that extensions to SysML can enforce clarity, consistency, and traceability, even in rapid development contexts where documentation traditionally lags behind implementation. By embedding verification attributes and quality constraints directly into model elements, teams can perform real-time validation against requirement baselines, ensuring that changes remain compliant with design intent. Complementary methodologies such as Model-Based Structured Requirements (MBSR) further strengthen this integration by linking model-based design, requirements management, and testing into a single cohesive architecture. Together, these frameworks operationalize MBSE principles within Agile's incremental cadence, ensuring that rapid development does not erode systems assurance.

The future of MBSE in Agile Software Systems Engineering lies in achieving a seamless balance between adaptability and engineering rigor. Methods such as SEFAD stand at the center of this transformation. As software increasingly defines the behavior of complex, safety-critical systems, traditional document-driven approaches cannot sustain the speed, traceability, and assurance required in modern development. SEFAD operationalizes MBSE within Agile environments by embedding structured modeling, digital requirements, and continuous verification directly into iterative workflows. This enables teams to maintain the responsiveness of Agile while ensuring every model, user story, and test trace back to validated system objectives. Through its tiered governance and model-driven reviews, SEFAD transforms MBSE from a static compliance tool into a dynamic, living architecture that evolves alongside the software itself. In doing so, it establishes the blueprint for future engineering practice, one where systems and software co-develop within a unified digital ecosystem of trust, traceability, and continuous assurance.

5.9 Conclusion

The integration of MBSE into Agile software development represents the most significant evolution in modern engineering practice, bridging the gap between rapid iteration and disciplined assurance. As software increasingly becomes the defining element of complex, mission-critical systems, the limitations of traditional document-centric engineering have become unsustainable. MBSE introduces a data-driven, model-centric environment where requirements, architecture, and verification coexist as synchronized, living entities. When fused with Agile's iterative and collaborative principles, this model-based paradigm enables organizations to accelerate delivery while maintaining the transparency, rigor, and traceability that high-assurance systems demand.

Frameworks such as SEFAD exemplify the operational future of this convergence. SEFAD transforms MBSE from a static design discipline into a continuous validation and assurance process, ensuring that models evolve dynamically alongside working software. Its tiered governance structure allows Agile teams to retain speed and flexibility while maintaining the systems integrity required by defense, aerospace, and other safety-critical domains. By embedding model-based verification and structured requirements directly into iterative workflows, SEFAD ensures that every user story, design update, and test execution remains grounded in an Authoritative Source of Truth.

The emergence of techniques such as MBSR, CMI, and digital thread architectures further advance this evolution, creating an ecosystem where system intent, model fidelity, and implementation progress are continuously aligned. These methods not only enhance communication and decision-making but also enable automated consistency checks, impact analysis, and real-time traceability across disciplines. In this environment, the models are engineering the system itself, no longer serving as supplementary documentation, directly increasing engineering and technical rigor.

Ultimately, the future of software systems engineering lies in the fusion of MBSE and Agile under frameworks like SEFAD. This approach unites adaptability with accountability, transforming how complex systems are conceived, developed, and sustained. The organizations that embrace this model will lead the next generation of digital engineering, transforming into one defined by continuous assurance, interconnected collaboration, and the seamless coevolution of models and code. Through MBSE-enabled Agile practices, the vision of a truly integrated digital engineering enterprise is not only achievable but inevitable.

5.10 Discussion Questions

Integration and balance: How does Model-Based Systems Engineering (MBSE) balance the need for engineering rigor with the flexibility demanded by Agile software development, particularly in safety-critical systems?

Cultural convergence: What cultural and procedural shifts must occur within engineering organizations to successfully merge MBSE's structure with Agile's iterative and collaborative mindset?

SEFAD's Significance: In what ways does the Systems Engineering Framework for Agile Development (SEFAD) operationalize MBSE principles, and why is it considered essential for the future of critical systems engineering?

Authoritative source of truth (ASoT): How does maintaining an Authoritative Source of Truth improve traceability and reduce risk in Agile environments, and what challenges might arise in sustaining it?

Model-based structured requirements (MBSR): How does MBSR enhance requirement quality and verification compared to traditional textual methods, and what implications does this have for Agile backlog refinement and sprint planning?

Digital Thread Impact: How does the digital thread transform lifecycle management and verification in Agile software systems engineering, and what risks exist if this thread becomes fragmented?

Sustainment and lifecycle evolution: How might the integration of MBSE and Agile under frameworks like SEFAD transform long-term system sustainment, modernization, and technical debt management across the software lifecycle?

Role of SysML: Why is SysML particularly suited for supporting MBSE within Agile contexts, and how does it strengthen communication among interdisciplinary teams?

Challenges in implementation: What are the most significant technical or organizational challenges in integrating MBSE into Agile workflows, and how might frameworks like SEFAD mitigate them?

Future vision: Considering the evolution toward Agile Model-Based Engineering (Agile-MBE) and Continuous Model Integration (CMI), what do you envision as the defining characteristics of future digital engineering ecosystems in critical system domains?

References

1. J. Wheaton, D.R. Herber, Digital requirements engineering with an INCOSE-derived SysML meta-model. in *Conference on Systems Engineering Research (CSER) 2024*, Tucson, AZ, 2024
2. D.R. Herber, K. Eftekhari-Shahroud, Building a requirements digital thread from concept to testing using model-based structured requirements applied to thrust reverser actuation system development. in *9th International Conference on Recent Advances in Aerospace Actuation Systems and Components*, Toulouse, France, 2023
3. Office of the Under Secretary of Defense for Research and Engineering, *Department of Defense Digital Engineering Strategy* (Department of Defense, Washington, 2018)
4. J.M. Marlowe, C.L. Haymes, P.L. Murphy, *NASA/TM–20220018538: NASA's Digital Transformation Strategic Framework & Implementation Approach* (NASA STI Program, Hampton, 2022)

Chapter 6
Case Studies in Implementation

Each software project is managed and executed in an individual and unique manner. Some projects follow a very rigid project management and systems engineering process, requiring formalized project documentation for every customer change, causing a large amount of effort on the project management and customer team, but leading to a very controlled development environment. These types of projects are quite common in the Department of Defense. It is more common for projects to be extremely Agile, reacting quickly to changing customer needs, but resulting in a very unstable development environment with increased project failure risk. In this chapter, we will show case studies discussing software projects using varying degrees of Predictive and Agile project management and engineering and technical rigor.

6.1 Case Study 1: Implementing Structured Requirements in a Defense Data Exchange Software System (DESS)

6.1.1 Background on the DESS System

The DoD Data Exchange Software System (DESS) is a data transfer system owned by Department of Defense. It is also utilized to host various other systems. Two teams perform sustainment of the system, a software development team and an information technology and cybersecurity team. A Branch Manager performs resource management for the teams with the project management performed by a Project Lead and the technical processes, policy, and system owned by a Systems Engineer.

Due to personnel issues, the branch responsible for the sustainment of the DESS system had to go without the oversight of a Systems Engineer for over a year.

D. Rosson, *Systems Engineering in Agile Software Development*,
https://doi.org/10.1007/978-3-032-14613-7_6

Without technical oversight, the duties were split up amongst other team members, primarily the Branch Head and Project Lead. Neither the Branch Head nor the Project Lead had formal training in systems engineering and struggled to build a good path forward for the project. Furthermore, the Project Lead had a background in IT and networking, with little to no experience in software development. This left the software team with no technical oversight. Due to the Wild West nature of this environment, most of the technical decisions fell upon the Software Development Lead, who struggled to plan, execute, and maintain system development goals while also attempting to train junior personnel. Most of the Software Development Lead's time was spent putting out fires as he was the only person who understood all aspects of the system. This further aggravated the situation as employees who should have been cross trained in the various subsystems, along with related status and development efforts, were required to pick up what they could from the little time the Software Development Lead had left in his day that could be allocated to them. In short, the DESS team had pockets of knowledge spread throughout the team, with no knowledge sharing occurring, high personnel turnover, and no overarching technical guidance. Due to these issues, the DESS team had not successfully released a major system update in over 2 years, with the last release occurring in March 2020. In March 2022, a Systems Engineer from another division in Keyport was rotated into the team temporarily to assist in solving the inability of the DESS team to successfully publish major releases and increase the technical rigor of the team.

Before the implementation of requirements management, the system was developed based upon verbal communication from the Program Office and through work prioritization as determined by the Software Development Lead. The team utilized a loose Agile Scrum methodology with work planned out on a Sprint-to-Sprint basis and rarely any further. There was no release plan in place nor a test plan. The development team worked on individual projects and the customer needs were met in a "who yells the loudest" fashion with special consideration given to Program Office requests. This led to an environment where large periods of time with no system updates were performed and in which the development team struggled to gain a cohesive vision of prioritized customer needs.

Testing in this environment was not thorough. With no requirements to test to, the development team relied on a procedure referred to in software development as a smoke test. Smoke tests are commonly used in software development as a method to ensure that changes to a codebase do not prevent the software from being launched. They are not to be used as formalized proof of testing. The smoke test listed out the areas of the DESS program and instructed the user to perform various data entries and form validation. It did not cover all areas of interest to the system and had poor coverage, allowing defects to be introduced as thorough testing was not taking place. Normally, a smoke test is used by a software developer prior to a software push to a repository to ensure that the code being pushed does not affect systems stability. It is a preliminary test and not to be considered a thorough test plan or utilized as a final testing solution with regards to verification of a system. Furthermore, with no test plan in place, the team found it exceedingly difficult to coordinate the complex production testing required for the environment the DESS system operated in. This environment included shore facilities on both coasts, as well as out of country, afloat vessels both in home port and docked outside of the continental United States (OCONUS), and internally at the Keyport facility. As is clear, the testing environment was complex, and adequate testing could not be accomplished in a single document smoke test as the team was attempting to do.

The lack of requirements led to the misunderstanding of potential changes to the DESS system. The change control board often argued over whether a task was a defect or new functionality. This made classification of work items essentially impossible to properly determine. All knowledge of what functionalities the system should have, along with the why it does the things it does and how it should perform those things, rested in tribal knowledge held by members of both management and the development team. This created a dynamic in the team where debates would commonly occur on whether a functionality was a defect and how the system should function in a given scenario. Most of the time, decisions were not made based upon noted customer needs, but rather the opinions of senior personnel. More junior personnel and those new to the teams were hard pressed to argue against said functionalities as without requirements documented, everything was pure guesswork and supposition. Furthermore, non-software development members of the change review board had truly little, if any, training in the difference between a defect and a change request, creating great confusion among what should have been a team with a full understanding of the development process.

6.1.2 Implementing Requirements Management

The first steps taken to move DESS into a systems engineering focused development methodology was to implement requirements management. A survey of the branch, starting with team leads and continuing down the ladder to the floor level technical personnel, determined that an extremely small percentage of the team had exposure to systems engineering style requirements management. There were hurdles to overcome, as it is common for software developers to push back against any method viewed as anti-Agile and a strong fear that they would be forced to transition to a Predictive, or Waterfall, methodology. The customer was not fully available for requirements generation and did not have knowledge that requirements were not being tracked. There were no requirements repository in place.

During the survey of the DESS team, it became readily apparent that there had been little exposure to systems engineering practices, let alone requirements management methodologies, by members of the team. Working closely with the Branch Head, time was set aside for training for the branch in requirements management. The team leaders were trained first with the plan to have the other teams trained after the leads had a good understanding of basic requirements engineering. As none of the team was engineers, this was the first exposure to requirements management methodologies that many had. As expected, it was a struggle at first, but the team readily accepted the changes, and a Requirements Management Board (RMB) was formed and stood up. The membership was to be the same members as the Government Leadership Team, including the Branch Head, Project Lead, Systems Engineer, Software Development Lead, Logistics Lead, Implementation Lead, and the Technical Writer. It was determined that the software was the first area that requirements needed to be generated for, and the Software Development Lead was tasked with working with the Systems Engineer to make it happen.

After the leadership team was trained, other teams were then introduced to requirements management, with most accepting the changes, though there was some pushback. The software team in particular was not open to change, having a fear of requirements management stemming from previous experiences in Waterfall style development. The major concern of the team was that if requirements were written down, there would be no flexibility in the development of the software, directly leading to a static development environment. The team wanted to still be able to continue their Agile Scrum style development, changing course as new needs arose. Careful consideration was given to their feelings, with one-on-one meetings held with developers that requested it and concerns addressed. By working closely with the development team, addressing their concerns, and hearing them out, the team was able to come to a point where they were open to accepting a change to a more rigorous requirements management methodology.

As no requirements repository was in place, the Systems Engineer was tasked with setting one up. The DESS team already utilized JIRA for project work tracking, so it was an easy choice to implement Requirements for JIRA (R4J) as the requirements repository. R4J is a plugin for JIRA that allows teams to create a top-down breakdown style tree of requirements, with full traceability to work items and test cases, along with obligatory parent–child relationships. It also allows for baselining of requirements and had a low entry point of learning how to use it as it functioned similarly to JIRA. R4J does lack in the ability to integrate with other systems modeling tools, such as CAMEO System Modeler, but the RMB decided that having the capability to link requirements directly to the existing JIRA work items was a much higher priority and decided to accept the lower functionality.

The software team was trained in systems engineering style requirements management. This included the utilization of "shall" statements and the importance of writing requirements that follow standard requirements practice as follows: "The subject shall do something under conditions within constraints/tolerances." The software development team struggled with the concrete nature of shall statements at first, oftentimes falling back on writing more vague requirements in a way similar to how they would write User Stories. User Stories are a way of writing software requirements in an abstract manner, and this team used the common method of the Connextra in the form of "As a {role}, I want {goal}, [so that{benefit}]." While User Stories are a fine tool for developing software, in a DoD environment where it is increasingly difficult to pivot and change system connections and needs, especially when dealing with the Submarine community, having rigidity in requirements definition is a positive and not a negative. Utilizing User Stories alone leaves too much interpretation of customer needs for the developer. By transitioning from User

Stories to Shall statements, the development team was able to start documenting the functionality of the system.

The development team started requirements documentation by working on each page of the DESS web application and subsystem separately. Starting with simple functionality, such as the login page, the team started to capture requirements and document them in a repository. For each page, a high-level requirement was generated. Children were populated under the high-level requirement and so forth, following standard requirements engineering practices. As the team generated requirements, they would periodically return to previous requirements, refining them, as necessary. The software developers quickly grew in confidence, learning what should be documented and what should not. The Systems Engineer functioned as a mentor and guide through the process, allowing the team to lead the generation and writing of requirements and stepping in only when needed to course correct or when advice was asked for.

As requirements were generated, change requests and defects started to be traced back to the requirements. The Configuration Management Plan (CMP) was revised to ensure that any new change request or defect must have an associated requirement, or it would not be worked on. A new workflow was devised and implemented, adding the RMB to the work approval process, as shown in Fig. 6.1. Through the implementation of requirements traceability, the leadership team was able to quickly determine the difference between a defect and a change request, something that the board struggled with prior to the more structured requirements management policy. This directly improved the DESS team's capability of identifying and prioritizing defects and change requests.

Within 9 months, the software team had completed enough requirements to perform an initial baseline. A method of determining the difference between new, approved, and depreciated requirements was needed. The team decided that new requirements were added with a metatag labeled as (NEW) until approved, at which point the metatag was removed. Approved requirements needed no changes. Depreciated requirements were marked with an (OBE) metatag, which stands for "Overcome by Events," a common term used in the DoD for depreciated or no longer needed items. This simple policy change allowed for the team to more easily search the R4J repository, writing Jira Query Language (JQL) queries to filter on those tags, or the lack thereof. It also allowed new change requests to be tied back to requirements, facilitating communication between the RMB and the CRB that a requirement had been generated for the change but had not been approved yet. This streamlined the process for the software team, giving them the power to document potential upcoming work without having to utilize a separate system to do so.

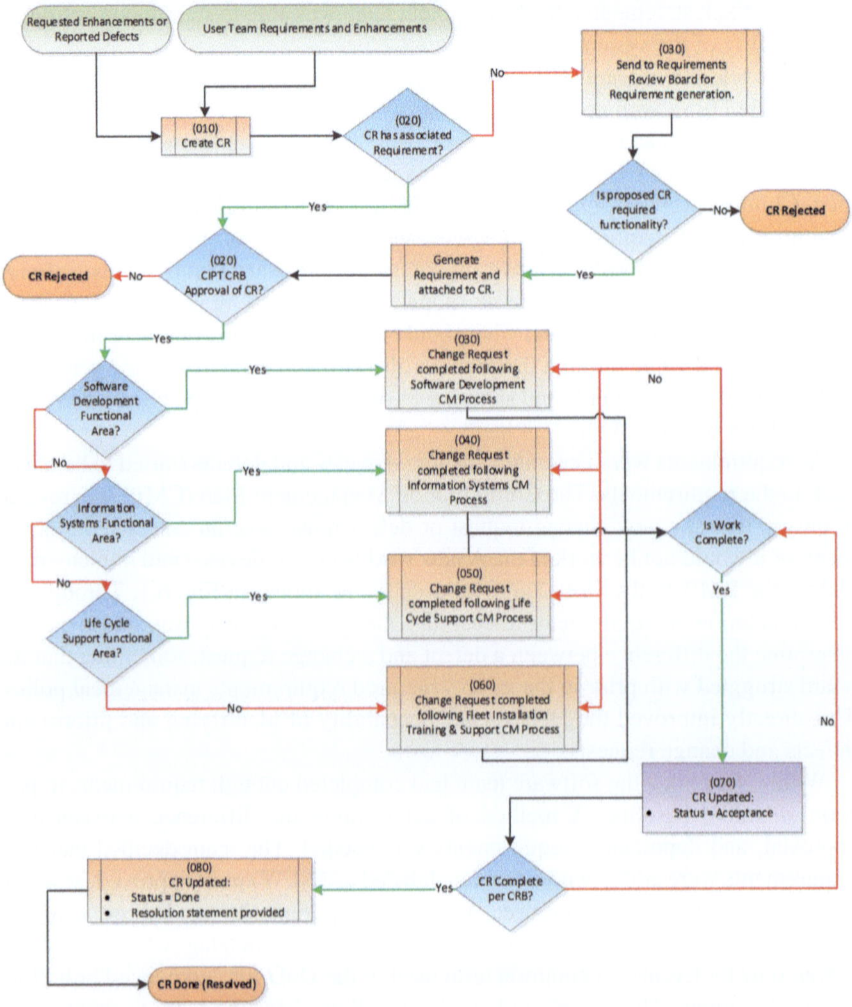

Fig. 6.1 DESS change request flow process

6.1.3 *Implementing Test Management*

As requirements were being developed, there was still the need for a formalized test management plan. As stated earlier, the DESS team had not successfully released an updated version in over 2 years. The sponsoring Program Office was not happy with this situation, as major changes to the system were tied to the updated version. Holding and passing a Test Readiness Review (TRR) was of paramount concern to the Systems Engineer and the team. In the DoD, a TRR "provides the formal approval authority with a review showing that the system is ready to enter the test and that the funding and execution of a test executes the test and gathers the required information" [1]. A successful TRR is required prior to a system completing formalized testing and being deemed ready for a Production Readiness Review (PRR).

The first steps taken by the Branch Manager and the Systems Engineer were to formally identify one of the development team members as a Test Engineer. While the person placed in the Test Engineer position did not have previous experience as a system tester, he did have a background and degree in Computer Science. To rectify the lack of knowledge, the Systems Engineer dedicated the first 2 months of the Test Engineer's time to training in systems test and evaluation (T&E). Time was set up with the DoD Logistics Management System (DLMS) Lead Test Engineer to assist in training the new tester, creating a mentor/mentee relationship. Once the training was complete, the Test Engineer had a much more solid foundation of T&E to work with.

As one of the key components of a test plan is having test cases, the team set about breaking the smoke test up into individual test cases. These cases were sorted, organized, and stored in the branch's Confluence site. Confluence allows files to show historical data, along with any changes, creating a configuration control system for test cases. This allowed tracking and security of the test cases, rather than storing them in a shared drive. As the smoke test was broken up, coverage was assessed. Where holes in coverage were noted, new test cases were created. This process continued until the DESS system test coverage was at 100%.

The Test Engineer would review requirements as they were posted in the R4J system. If a requirement represented new functionality that was not previously accounted for in the test cases, a new case would be written and documented. The Test Engineer would remove functionality from the test cases if the requirement were for an OBE function. This created a system that accounted for changes in requirements, directly saving time and money for the team by ensuring the test plan was kept up to date.

As formal test cases were being developed, the system still required a major release to be executed. To facilitate the DESS 4.1 release that had been in the queue for 2 years, the Systems Engineer devised a plan to ensure all discrete system types were tested. This required facilitation of testing aboard afloat commands, both CONUS and OCONUS, as well as various shore facilities. It was determined that to have full coverage for the fleet, SSNs would be identified with connections both via the Navy & Marine Corps Intranet (defense) and through non-defense connectivity. SSBNs would also have to be tested, but only through a pier connection. The Type Commander (TYCOM), SSBN Offcrew offices and Trident Training Facilities and various other local assets would also have to participate in testing.

As each platform utilizing DESS had different requirements associated with them, a test matrix was devised that showed test coverage. The smoke test was still available, along with the DESS User Guide. The User Guide detailed step by step instructions to perform specific functionalities. It was utilized by Sailors to perform their duties and to learn DESS. As the Sailors were familiar with the User Guide, it was determined that only internal testing would utilize the new test cases and external testing would be performed following steps in the User Guide. This was communicated and approved through the Program Office and TYCOM. A matrix was devised that tracked what and how each platform would test, as shown in Table 6.1.

By June 2022, the team had gotten approval to pass through the TRR. They proceeded to conduct a full systems test, successfully ensuring that all parties performed systems testing as needed. Prior to publication of the updated version, the DESS system had to successfully gain approval for release to production at a PRR. For DoD programs, a PRR "determines whether the system design is ready for production, and whether the developer has accomplished adequate production planning for entering Low-Rate Initial Production (LRIP) and Full-Rate Production (FRP)" [2]. As of In July 2022, DESS v4.1.1 was able to successfully hold and pass a PRR, gaining approval from the Program Office for distribution to the fleet. Gold disc production was scheduled, as well as the concurrent release of an updated User Guide.

6.1.4 Results of Policy Changes in DESS

Applying requirements management principles and formalized test management to the DESS system resulted in increased efficiency and team cohesiveness. Through the utilization of requirements management versus loose User Story tracking, the development team increased their knowledge of what the system was required to do and how it should function. This directly created an environment where a formal test plan could be created and tracked. Furthermore, the difference between change requests and defects were solidified. Utilizing R4J allowed the team to trace all changes and defects directly to requirements, easing the ability to note new functionality from existing.

Of special note is the working environment and stress level of the software development team. Before the implementation of the requirements management and formal test planning, the team often reported high stress levels and frequent confusion on what needed to be worked on. Few had a full understanding of what the full system should do and how it should behave in most scenarios. Knowledge was very pocketed with most knowledge transfer occurring on the job and large swaths of information held in tribal knowledge. After the changes to process and policy were enacted, the team frequently reported higher job satisfaction, less stress, and increased understanding of the system as a whole. While this is just one team, it is a good example of how bringing systems engineering methodologies into Agile Scrum environments can be greatly beneficial to programs.

Table 6.1 DESS v4.1.1 test matrix

Test plan Release version: 4.1.1 Date: 5/10/2022 Matrix version: 1.0		Tester group				
		Fleet testers	Fleet testers (special)	TYCOM	Local	Total tests required/passed
Required		6	6	14	21	47
User guide						
4.1.1	Afloat to ashore transfer process	x				
4.2.1	Data transfer: Backup (afloat)	x				
4.2.2	Data transfer: Restore (ashore)	x				
Production testing						
	LAN admin functionalities					
	Login			x	x	
	Home page			x	x	
	LAN admin landing page			x	x	
	Backup process			x	x	
	Restore process			x	x	
	Account management			x	x	
	Download/upload			x	x	
	User functionality					
	Login			x	x	
	Home page			x	x	
	Third party application access			x	x	
	Super admin functionalities					
	Login				x	
	Home page				x	
	Super admin dashboard				x	
	Configuration control				x	
	Crew management				x	
	Link configurations				x	
	Announcements				x	
	Third-party application access					
	Afloat readiness tracker	x	x	x	x	
	Afloat Onboard training			x	x	
	Training manual tracker	x	x	x	x	
	Downloads			x	x	
Fleet access test						
	Local area admin access					

(continued)

Table 6.1 (continued)

Test plan Release version: 4.1.1 Date: 5/10/2022 Matrix version: 1.0		Tester group				
		Fleet testers	Fleet testers (special)	TYCOM	Local	Total tests required/passed
	Login and landing page		x			
	LAN admin view		x			
	User access					
	Login and landing page		x			
	Upload support					
	Upload test	x				

6.2 Case Study 2: Implementing Systems Engineering Rigor in a Defense Logistics Management System (DLMS)

6.2.1 *Introduction*

Many software projects in the Department of Defense (DoD) lack the benefits of established systems engineering methodologies and good engineering rigor. Software developers are trained in the use of the various Agile frameworks and have extraordinarily little exposure to systems engineering methodologies. It is common for software developers to be strongly opposed to any development methodology that can be viewed as infringing on the principles of the Agile Manifesto. Only large-scale DoD software projects executed under the direction of the DoD Instruction 5000.02, Operation of the Adaptive Acquisition Framework, are required to follow rigorous systems engineering methods [3]. Due to this, many smaller projects in the DoD have little engineering rigor and frequently fail to meet customer needs and expectations. The DoD Logistics Management System (DLMS) is one such project. The following text details the transition of the DLMS project development from no engineering rigor to one that merges systems engineering methodologies with Agile Scrum principles. By selectively increasing engineering rigor through merging systems engineering principles while maintaining and leveraging the flexibility of the Agile Scrum Framework in the DLMS project, the project lowered risks associated with cost and schedule while increasing customer collaboration and communication.

6.2.2 Background on the DLMS Project

The DoD Logistics Management System is a Diminishing Manufacturing Sources and Material Shortages (DMSMS)[1] management system that facilitates management of obsolescence of equipment and parts which is developed, maintained, and utilized by the Department of Defense. The DLMS system is used to determine mitigation strategies, manage military platforms, such as ships, aircraft, and land vehicles, and perform logistics and lifecycle planning. The DLMS system started as a small system and has grown into a large, multifunctional system. This growth took place with little thought of systems engineering, engineering rigor, or technical rigor.

DLMS, created in the early 2010s, evolved from a spreadsheet-based system into a web application. As the spreadsheet system was used, it was quickly determined that a centralized data repository in a web-based system was necessary. A software development firm was contracted to transition the spreadsheet-based system into a web application, which was then turned over to the DoD for sustainment. A small software development team consisting of one government and one contractor was created to maintain the system and develop new capabilities.

Though small, the development team was made up of senior programmers that had been developing software since the late 80s and early 90s, and the team utilized their own form of project management. Neither developer had a formal education in software development, having been fully self-taught during the dot-com bubble. The team used a SharePoint system to track proposed changes and defects. Changes requests were referred to as "enhancements" and written in the form of User Stories, a practice that originated in Extreme Programming practices. User Stories are normally written utilizing the Connextra template, which is as follows: "As a {role}, I want {goal}, [so that{benefit}]." Both defects and enhancements were prioritized using a high, medium, and low priority system, though this system was not clearly defined. The team did not plan work based upon Sprints or utilize a work tracking system, such as a Kanban board. Each team member pulled work from the backlog that they felt needed to be done and accomplished it at their own pace. There was no plan for code publishes to the production system and testing was performed through a large "smoke test." A smoke test is a general test procedure that is performed to verify the main functionality of a software system, traditionally used to ensure that changes made during development will not prevent the software from launching. The DLMS smoke test was a large test case that walked through the core functionality of the system but did not cover all edge cases nor was it designed to test individual subsystems or functions. Engineering rigor was low to non-existent, and the

[1] DMSMS is a field of logistics management that is concerned with the obsolescence of parts, raw materials, or software when the companies that produce them discontinue their manufacture, leading to shortages and potential supply chain disruptions.

project struggled to meet Capability Maturity Model Integration (CMMI)[2] standards as set forth by the department. Over the course of 5 years, the team grew to five developers, including the addition of a computer scientist, a database administrator, and a test engineer.

The DLMS system oversight is the responsibility of a governing body called the Stakeholder Review Board (SRB). During this time, SRB membership was made up of each of the Branch Managers in the division, the Division Head, Technical Project Managers with projects associated to DLMS, and the DLMS Project Lead. The SRB acts as a steering committee, making directional decisions for the system, approving funds appropriation, and overseeing project execution.

The SRB delegated the authority to approve change requests to the DLMS system through another committee called the Enhancement Review Board (ERB), which was comprised of the SRB plus the addition of the other divisional Project Leads and DLMS power users. The ERB had the authority to approve or deny enhancement requests. This created a close working relationship with customers where customers had a large sense of ownership in the DLMS system.

6.2.3 Identifying there Is a Problem

At this point, numerous new functionalities were developed in the DLMS system that the user base did not desire. As the development team tracked enhancements utilizing User Stories, little detail of what the customer needed was communicated to the developers. Instead, the individual developer met with the enhancement proposer to figure out what the customer desired. This collected data was normally kept handwritten in notebooks, which were scattered about the workspace. There was no electronic record or notetaking associated with the development. This led to a system where the developer thought they understood what the customer desired but did not have a standardized way of documenting this or communicating it back to the customer. No modeling was performed. SharePoint tickets were updated but no tracking system was used.

Due to the lack of formal work tracking and communication methods, customers rarely received the changes or projects as desired. As no estimation tool was utilized, it was common for customers to hear that it would take "2 weeks" to complete a change, regardless the size or complexity, so much so that it became a running joke within the customer group. Change requests routinely were delivered with functionality that was not desired by the customer or that did not quite meet the requirements of the customer needs. With no formal work tracking or system for workload estimation, developers were left to their own devices, often spending time working on pet projects or "fun little tweaks" to the system, adding new color

[2] CMMI is a framework that is used in industry to gauge and improve process maturity in organizations. It categorizes processes into five levels: initial, managed, defined, quantitatively managed, and optimizing. See more at https://cmmiinstitute.com/.

schemes for the interface or small changes to the system that customers did not request.

At this time, projects often started without consistent User collaboration and with little understanding of the proposed system context. An example of this is the DLMS Watch List functionality. The Users desired to be able to track risk status of various parts, adding them to a watch list style function. In DoD logistics, watch lists are a report that captures the status, mitigation plans, and trends risks that must be elevated to senior leaders. Customers attempted to communicate this idea to the DLMS development team in a series of meetings. The team did not understand the problem domain and instead came to the assumption that the Users desired to be able to create a list of parts to be flagged for follow up on a scheduled basis. A function in the system was developed allowing users to mark parts with a flag, a recurrence period, and criticality. These parts were then viewed in two methods. The first was as a pop-up reminder upon logging into the DLMS system. The second was through a menu item displaying a list of all parts with their associated watch list meta data. During the development of the DLMS Watch List, Users were not consulted past the initial meetings nor were any prototypes demonstrated. When the functionality was finally presented to the requesting customers, one User stated, "I could see how this could be useful, but this isn't anywhere what I had in mind." Due to a failure in communication resulting in a misunderstanding of the basic needs of the customer, a subsystem was designed and implemented into the DLMS system that was not desired. It took another year of tweaks in functionality through User input to make the watch list perform the developers intended function. Users never received the actual logistics watch list functionality they desired.

Within 6 months of joining the team, it was clear that some form of software development rigor was required. Customer dissatisfaction was high, and projects were not delivered within cost or schedule. A system was put in place to incorporate "User Design Teams" (UDTs) into the development process. UDTs consisted of users of the DLMS system representing each of the branches of the division with a member of the software development team facilitating the meeting. The UDT members would sign a charter for the associated project and agree to availability for project reviews and to answer questions of the development team.

6.2.4 Applying the User Design Team Process

Utilizing the UDT process, the team increased the success rate of project delivery. Work items associated with projects were now tracked in a formal system, as User Stories kept in a spreadsheet. Engineering rigor was still low, as there was no modeling, no complexity or workload tracking, and little schedule projection. However, customer engagement was at an all-time high and satisfaction with the software product was increasing. While the User Stories were not detailed enough to capture all the customer requirements, they were more detailed and tracked than prior to the UDT process.

Issues still occur using UDTs, mostly due to a lack of communication and understanding of customer needs. An example project where this occurred is the Supportability and Forecasting Tool (SFT) subsystem. The SFT started as a spreadsheet utilized by an obsolescence management group in the DoD. It used the built-in functionality of Excel to leverage the Poisson distribution, allowing logisticians to forecast part availability and assist in identifying risks associated with the supply chain. This was tied directly to submarine schedules, based upon hull designations. The spreadsheet had the full functionality desired by the logisticians but lacked a live connection to the DLMS database. Data was stored over multiple spreadsheets, causing issues between logisticians. Customers desired to transition spreadsheet functionality to the DLMS system, allowing for a common workspace for the SFT users.

The development team used the UDT process to develop the SFT. A UDT team consisting of users of the SFT spreadsheet was assembled and a charter was signed. The team worked together to create User Stories to represent the SFT system. The spreadsheet was given to the development team to reverse engineer. Complexity was high in the system, as discrete probability curves had to be calculated utilizing the Poisson method and the limited resources with experience in discrete mathematics. The logisticians that utilized the SFT spreadsheet did not understand the math behind what they were using, relying on common logistics knowledge to know to apply the Poisson distribution.

The project was successfully delivered on time and within budget but with a limited number of the functionalities that the users required. The limitations of User Stories and a lack of requirements traceability with no task planning led to many assumptions being made by the development team. The product looked and functioned similarly to the spreadsheet based SFT but was not user friendly. The required schedule tool was hard for users to use and had no way to automate or upload schedules, requiring users to manually input dates and hulls. Due to limits in the coding language used and the hardware which DLMS was hosted, open-source software packages were required to calculate the Poisson distribution to a level that did not cause crashes in the system. This required the developers to go line by line through the software, purging out what they felt was unnecessary. Even with the purges, the system ran much slower than the Excel spreadsheet version and took longer to input and manage data. This resulted in the customer not using the DLMS based system and instead continuing the use of the spreadsheet version. Five years later, the test team found that not only had zero users logged into the SFT system in the last 2 years, but that the major functionality had ceased to operate due to changes in the code base versions. With permission from the Stakeholder Review Board, the SFT subsystem was purged from DLMS.

6.2.5 Transitioning to Agile Scrum

The development team continued to grow, adding four more developers. With a larger group, it was determined that the team needed to work in a more organized manner, as schedule slippages were common and there was no visibility into what each individual developer was working on. It was decided that the team would transition to utilizing the Agile Scrum Framework.

As required by Agile Scrum, one of the developers was assigned as the Scrum Master. The Scrum Master's duties were to write the DLMS Scrum Handbook, manage the Product and Sprint Backlogs, and train the development and customer teams in the Scrum methodology. A large whiteboard was assigned as the Sprint Backlog for the team and an area cleared out around it so that the team could hold daily Scrums in front of the board. The Product Backlog was still managed in the same SharePoint as before, but Story Points were now assigned to work items. Story Points are not part of the core of the Agile Scrum Methodology but instead are drawn from the eXtreme Programming method and are commonly utilized by Scrum Teams to assist in estimating workloads for Sprints and Product Increments. The Scrum Master started to track Sprint Velocity and held Backlog Grooming and Sprint Planning meetings. Sprint Reviews were held with stakeholders and customers invited. The Scrum Master would facilitate Sprint Retrospectives so that the team could discuss how the previous Sprint went, identifying and integrating the "good" aspects of the last Sprint into the Scrum and development policies and determining how to avoid the "bad" in future executions.

The DLMS Project Lead was assigned as the Scrum Product Owner role. He took on the responsibility of representing customer interests, accepting work completed during Sprint Reviews, assisting customers during UDTs, and ensuring that the Product Backlog was well formed, managed, and the User Stories were within compliance of the Connextra method. He stopped working as a front-end developer and instead focused his development time to work on research projects with the University of Washington and perform database administration and development, reducing his impact on Sprint execution.

The development team worked to bring more clarity into work items associated with Sprints. During Backlog Grooming, the team would break down large Story Point User Stories into smaller ones that they felt were more easily accomplished. Any story with a value over 40 Story Points was flagged as an Epic story. Once a User Story had been adequately reduced to a value that a team member could accomplish within a day or two, Tasks were associated with the Story. Tasks were written in plain English and did not follow the Connextra format. This created a system with more traceability than was present by utilizing only User Stories to develop new features and functionalities. This was accomplished through Epics being broken down into User Stories which were then broken down into Tasks. User Stories and associated Tasks were printed out on sticky notes and put on display on the Sprint Board. Status was denoted by the column in which the sticky note was present. The status columns were Backlog, In Process, Ready for Testing, and Done.

Developers claiming a work item signed the sticky note, taking ownership of the work item.

By tracking Velocity and work item complexity through Story Points, the team was better able to plan how long each feature effort would take to complete. This allowed customers to make better decisions on the prioritization of software enhancements, creating a more organized backlog with a clear structure for customer needs. The Project Lead was better able to plan out how much work could be accomplished in a fiscal year and how many developers were needed per effort. This brought more rigor into the engineering and project management efforts.

During this time, an effort was made to create a more formal and rigorous test management plan. The smoke test was divided into larger sections, demarcating the various subsystems of DLMS. The test engineer then went into each subsystem, detailing functionalities in an "as built" manner and noting them in test cases. These test cases were then assembled into a spreadsheet to keep track of them. Prior to each production publication, the test cases were walked through and signed off by the test engineer and the development team, ensuring that if a developer had worked on the subsystem, they would not be a tester on that specific page. While the broken-out test cases were a vast improvement over the single smoke test, there was still a distinct lack of traceability to requirements, which created an environment of operating by tribal knowledge and arguments with customers over what functionalities should be present and why. Though not traced to requirements, the breakout of individual test cases from the smoke test increased the capability and confidence of the team in delivering a working product to the customer.

6.2.6 Moving to a Requirements-Driven Development Method

The DLMS team continued to grow, transitioning to be housed in their own branch in the division. With the move to a new branch came new leadership in the form of a new Branch Manager. The Branch Manager had previous work as a systems engineer, with expectations of more engineering rigor for the DLMS project. A new team of contractors also joined the branch, performing commercial off-the-shelf (COTS) vendor survey activities. The team was now at 30 members, having grown from the previous five members over the course of a 5-year period.

With rapid team size growth, the need for structure became increasingly necessary. The previous Project Lead and Product Owner transitioned to be the Staff Engineer. The Scrum Master transitioned to the Project Lead and Product Owner position. The rest of the staff were broken into two teams, the Development Team, and the Survey Team, with a Scrum Master assigned to each one and sub-teams having assigned Technical Team Leads. With the branch structured in a more hierarchical manner, decisions were easier to manage and team members better understood individual responsibilities.

The introduction of the Staff Engineer allowed the team to start the transition to a requirements-driven software development methodology. The role of the Staff

Engineer, filling the role traditionally staffed by a Systems Engineer, included ensuring that customers and stakeholders were directly involved in requirements definition and analysis, Model Based Systems Engineering activities, including architecture definition, and verification and validation activities. As the Staff Engineer was close to retirement, the Product Owner pursued formal education in Systems Engineering. The Staff Engineer and Product Owner drove the team toward implementing systems engineering principles, including requirements management.

The team also used this time to implement better knowledge management and project management controls. Instances of JIRA, Confluence, and Git were stood up, with use policy documented. The Product Backlog was transferred from SharePoint to JIRA. The team used the built in JIRA functionality to manage Sprints and Sprint Planning. At first, the team used both a physical Scrum board and the digital JIRA Scrum board, but eventually moved solely to the JIRA Scrum board. While this took adaptation, it did allow for more knowledge management with regards to ticket history as developers were able to more easily annotate comments and pertinent files to the JIRA ticket items. Customers were also able to easily comment and provide feedback while watching progress status on enhancements and defects that they submitted, giving a level of visibility into the development process that was obfuscated prior.

The ERB was transitioned to a Change Request Review Board (CRRB). Enhancements were renamed to change requests to be more in line with industry standards. The CRRB functioned very similarly to the way the ERB did, but the membership was reduced to sole users. Each customer branch was allocated two voting seats and for a CRRB meeting to be held, at least one member from each branch must be present. The COTS Survey Team Lead held a voting seat representing the Vendor Survey efforts. The Product Owner was in attendance, and the Scrum Master facilitated the meeting. The CRRB took greater responsibility in the ownership of the change requests, having the authority to not only approve or turn down, but also being the only personnel allowed to submit change requests. All change requests were submitted directly through branch CRRB members, ensuring that change requests were written in a consistent manner. The CRRB also had the authority to determine the priority of the backlog of change requests, creating a "top 10" system where the change requests they wanted worked on would be marked as High priority, those that were in queue to be worked on in the future marked as Medium priority, and those that were considered too low for scheduling but to be done if the opportunity presented itself, referred to as "low hanging fruit", were marked as Low priority. By having the customer approve and prioritize proposed changes to the system, the development team directly benefited from knowing exactly what the customers wanted. The customers also had to directly weigh the pros and cons of the impact on development for each change request, leading to higher quality and more critical changes being approved.

A repository in JIRA was stood up to hold requirements. The team utilized the R4J plugin, allowing for ease of traceability from a requirement to a JIRA work item, as it was readily accessible to them and a no cost solution. Using R4J increased understanding of work items to the team members and allowed the Test Engineer to better understand the requirements and the traceability to test procedures.

For requirements approval, the team approached the Stakeholder Review Board (SRB) for direction. Being primarily made up of management, the SRB decided to delegate requirements approval to a new board, the Requirements Review Board (RRB). The membership of the RRB was initially made up of the same membership of the CRRB, as the CRRB was seen to have the most knowledgeable users in the division with regards to DMSMS and DLMS use. The RRB held the authority to approve, deny, or modify requirements for the DLMS system, with the SRB retaining veto authority and tie breaker vote if needed.

With the repository in place and the development team trained in basic requirements generation and management, the team proceeded to the task of converting functionalities present in each subsystem of DLMS to an "as built" requirements document. Twenty-three functional areas were initially identified, with more being added as the effort was executed, ending up with over 50 functional areas. Examples of the functional areas can be seen in Fig. 6.2.

INCOSE requirement standards were followed with requirements written in system engineering requirements management shall statement format. This was decided as most of the customers that the developers interacted with were either engineers or logisticians. Engineers are trained in shall statement style requirements management and many logisticians have at least a surface level understanding of the style. INCOSE standard shall statements are also easily understood, using a parent–child traceability and breakdown structure, which makes the format quite easy to train personnel in. Customers can be trained in writing Connextra-style User Stories, but DLMS customers found that shall statements were much more familiar and easier to communicate their needs. See Fig. 6.3 for an example of how these requirements were written and stored in R4J.

The team used a common spreadsheet template to document the requirements. A tracking spreadsheet was developed to note the status of each functional area with the following information tracked: Requirement Area, Requirement Numbering, Sub-Requirements, Assigned To, Needs Reworked (Y/N), Ready for Review (Y/N), Initial Review Done (Y/N), Requirements Review Board Approved (Y/N), Uploaded to R4J (Y/N), Notes, Time to Generate (in weeks), Time for Initial Review (in hours), Time for Final Review (in hours), Number of Requirements (see Fig. 6.4). This gave a good insight from a project management perspective into the amount of effort required to develop the as built requirement set and as to how much more effort would be required to complete the set and create an initial baseline.

While in the process of creating the As Built baseline requirements set, the project team still had the responsibility to continue previously scheduled projects. To accomplish this while maintaining good requirements management practices, the development team ensured that all new projects had baseline requirements generated prior to the execution of the project. This required some work to fully understand how to accomplish this, as normally requirements are very rigid. Many times, customers view requirements as something that is chiseled in stone and that should not change. Not wishing to give up too much of their Agile mindset and execution

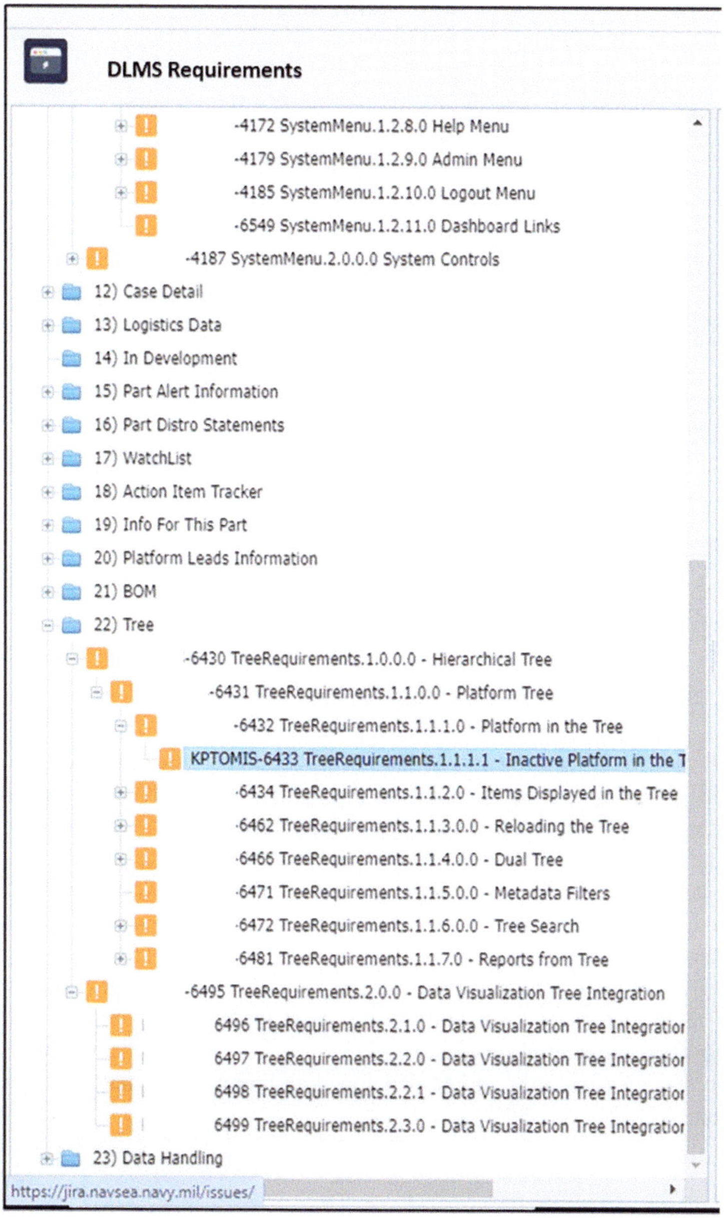

Fig. 6.2 Example of DLMS functional areas and requirements structure

policies, the development team treated requirements sets much as they would a set of User Stories. They were prioritized and linked directly to User Stories for Story Point allocation. This allowed the team to better estimate the level of effort required to accomplish a requirement. The team also frequently revisited requirements with

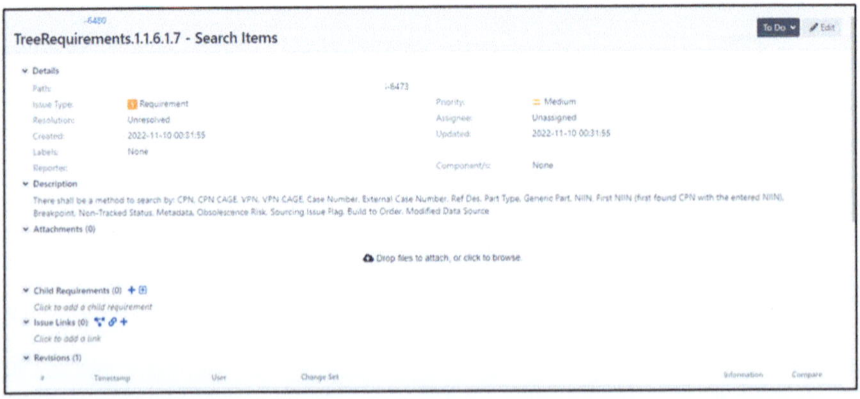

Fig. 6.3 DLMS requirements sample

Requirement Area	Requirement Numbering	Sub Requirements	Assigned	Needs Rework	Ready for Review	Initial Review Done	RRB	Uploaded R4I	Notes	Time to Generate (in weeks)	Time for Initial Review (in hours)	Time for Final Review (in hours)	Number of Reqs
Data Dashboard	DataDashboard.x.x.x		Bob	Y	Y	N	N	N	Security Updates	0	0	0	107
Login	Login.x.x.x		Sally	N	Y	Y	Y	Y		0	0	0	13
Part Detail	PartDetail.x.x.x		Christian	N	Y	Y	Y	Y		0	2.5	0	120
Reports	Reports.x.x.x	Admin, Customer, Metrics, Utility	Sally	Y	Y	N	N	N		6	6		400
RC2	RC2.x.x.x		Bob	N	Y	Y	N	N		0.5	0.5		25
Vendor Survey	VendorSurvey.x.x.x	Conduct Survey, Performance, Load Balancer, Log Viewer, Group Management	Christian / Frank	N	Y	Y	Y	Y		0	0	0	130
RC1	RC1.x.x.x		Christian	N	Y	Y	Y	Y		1	1		61
Admin	Admin.x.x.x	Import External Data, Manage Application, Security Management, Logs, Developer Help Files, Admin Mode	Daniel	N	Y	Y	Y	Y		1	1		50
System Menus	SystemMenus.x.x.x	System Menus and Controls	Daniel	N	Y	Y	Y	Y		0.5	1		50

Fig. 6.4 DLMS as-built functional area requirements tracking

the SRB and RRB for projects, adding, removing, and adjusting requirements as necessary to reflect the current state of the software development effort and customer desires. A true baseline was not snapped until the project was considered finished, final delivery was made, and the documentation closed out.

The development team members and customers responded very well to the increased engineering rigor associated with good requirements management. Customers reported increased satisfaction with the development efforts of the DLMS team. Having a clearer understanding of customer needs allowed the development team to focus on creating exactly what the customer wanted rather than focusing on non-essential but perceived desires. The test engineer found it much easier to execute his duties and he developed an increasingly mature test management plan.

6.2.7 Utilizing Wireframe Models to Facilitate Communication

To facilitate further communication with the customer, the development team started utilizing models during presentations in the form of rapid prototyping. Rapid prototyping is the use of creating models, drawings, or sketches to assist in determining how a User will utilize a user interface. The developers took customer requirements and transitioned those into mockups of how they envisioned creating the system pages. This allowed customers to better understand what the development team planned and suggest adjustments if needed while also engaging customers earlier in the design process. Paper prototyping activities also took place where a User walked through menu options and use cases with the models. This directly increased the developers understanding of how the User would use the page. Rapid prototyping using wireframe models in this method is highly effective with regards to the small amount of effort required to perform the process and the high rate of return. It also often leads to increases in developer and customer communication. Wireframes are visual models widely used to represent graphic user interfaces and system functionality in Agile development.

The development team found modeling to be especially useful in rapid prototyping "widgets" for the DLMS dashboard. The DLMS dashboard utilized a sandbox area where Users could choose various small widgets to display data. These widgets were designed in such a way that a software developer could create one within a few hours if given the proper data. Combining widgets together in a dashboard view allowed customers to create custom reports and dashboards without having to request a formal report through the change request process. Customers also found this approach helpful in communicating their needs and often would create their own wireframe models to present their ideas to the development team. An example of a model of a widget can be seen in Fig. 6.5.

While rapid prototyping activities utilizing wireframe models is not strictly a Model Based Systems Engineering (MBSE) activity, it laid the foundation for the DLMS team to start utilizing MBSE methodologies in their software development. As customers became more comfortable with the idea of seeing designs laid out in models, the team utilized other techniques, such as SYSML and the Department of Defense Architectural Framework (DoDAF), to model out future projects. An example of this can be seen in Fig. 6.6, an OV-1 DoDAF view created for a proposed Commercial Off the Shelf Part Selection Tool. Models such as this increase understanding and communication between customers and developers, leading to lowered risks to cost, schedule, and performance through increased understanding of the problem domain.

6.2.8 Results of Implementing Systems Engineering Rigor in the DLMS System

Numerous positive changes resulted from implementing systems engineering rigor into the DLMS development lifecycle. The DLMS development team had a much easier time understanding customer needs and expectations. Software defects have been reduced in the project and test personnel have a clearer understanding of why and how the software should operate. Reported customer satisfaction is at an

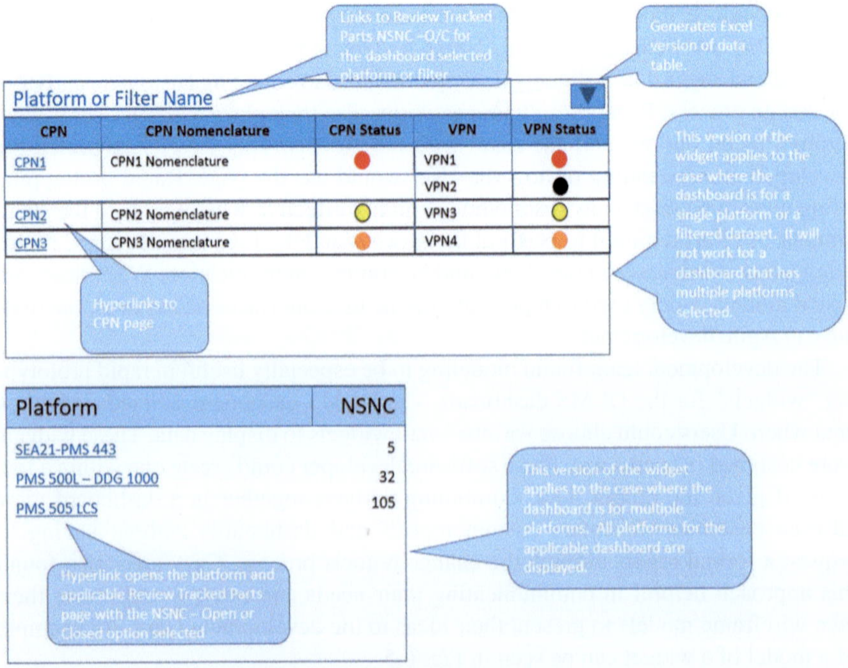

Fig. 6.5 Wireframe model of a DLMS widget

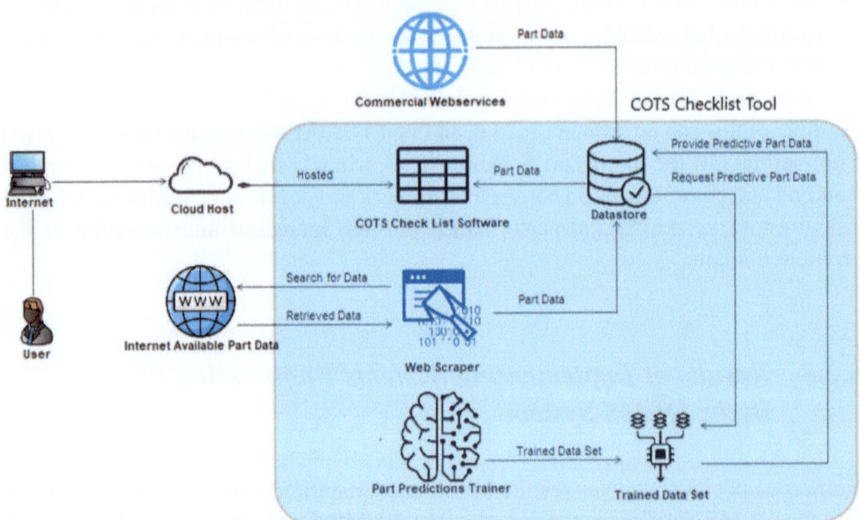

Fig. 6.6 COTS selection tool OV-1 example

all-time high. Overall, the DLMS project is in a more improved state than it was prior to implementing the above changes.

The DLMS development team has experienced an increase in customer communication and collaboration, as well as reporting a better understanding of customer needs and expectations. Developers state that Systems Engineering Requirements Management shall statements "paint a more detailed and clearer picture of what customers want" than when the team was solely utilizing user stories. By using MBSE techniques, such as modeling new functionality prior to coding efforts, developers can present the functionality in a rapid and cost-efficient manner to customers, gaining much needed feedback during the initial development steps, rather than waiting for sprint reviews to get customer feedback. The team enjoys the collaborative environment this method has created, embracing customer created wireframes and drawings as system development enablers. Utilizing a more rigorous engineering foundation has resulted in improved developer job satisfaction, a higher understanding of customer needs, and a more collaborative working environment.

Defects are costly for any system, including DLMS. Therefore, the test team reports the greatest increase in job satisfaction from implementing systems engineering requirements management policies. Having to rely on a single, large smoke test resulted in many defects being published to the DLMS production environment. Though the work was arduous to create an "As Built" baseline of requirements, the test team reports much greater ease in tracing customer needs and system requirements to test cases, resulting in a rigorous and robust test management plan and process. Requirements are traced from inception to work item, allowing the test team to know exactly what needs to be tested, why it needs to be tested, and what the results of testing should look like. Defects in the DLMS Product Backlog have decreased by 300% in comparison to prior to implementing Requirements Management. The change to a Requirements Management centric software development method has directly increased the ability of the test team to meet test management requirements and ensure engineering rigor.

DLMS customers report a record high in satisfaction with system and project interaction. The DLMS Project Lead conducted a survey of customers in January 2024. System users and technical managers were asked a series of questions, and the responses were collated. The results of the survey are displayed in Table 6.2.

The implementation of systems engineering rigor into the DLMS software development lifecycle created a dramatic positive effect on the project. The development team has increased throughput of their development product, having increased confidence in customer needs and desires. The DLMS Product Backlog has seen a reduction in defects by a remarkable 300%. The test engineer and development team report reduced rework from customer dissatisfaction and a reduction in unmet needs. Surveyed customers have reported much higher satisfaction with the product and an increased feeling of ownership in the system, with respondents averaging a score of 8 out of 10, resulting in a "high" satisfaction rating. Overall, implementing requirements management, system modeling, and stringent test planning and management has had a positive impact on the DLMS system.

Table 6.2 2024 DLMS customer satisfaction survey results

Question	Average result	Notes
Prefer user stories or shall statements?	Shall statements	83% prefer shall statements 17% prefer user stories
Are you more or less confident in the quality of the DLMS system?	Equal/more	50% as confident 50% more confident
Do you feel prototypes or wireframes have a positive impact in understanding and communicating user needs?	Yes	100% positive impact
In the last 2 years, have you felt you are finding more or less bugs/defects in the system than in years prior?	Less	100% feel less bugs/defects
Overall, how satisfied are you with the DLMS development process as opposed to years prior? (on a scale of 1–10, 1 being unsatisfied and 10 being very satisfied)	8	Average calculated based upon all respondents
Do you feel communication and development status has become more transparent or less in the last 2 years?	More transparent	67% improved communication, 33% equal to previous years

6.3 Case Study 3: Project History of the DoD Process Automation System (DPAS) 3.0 Client/Server to Web Application Server Conversion Project

6.3.1 Introduction

The DoD Process Automation System is a process control system built and sustained by the Department of Defense. DPAS was not the first process control system the DoD developed and sustained. In 1986, the DoD deployed a previous inventory and document system referred to as the DoD Control System (DCS). After 10 years in service, DoD leadership recognized the need for a more robust system than DCS could offer and, in 1999, fielded the DPAS 1.0 system.

DPAS was deployed as a three-tier client/server-based system, as was common at the time [4]. Each system build was managed by the development team and pushed from the server to the client. For a defense seat to have DPAS installed, the DPAS software system must be manually installed by a system administrator. Once the DPAS system is installed, client updates are checked each time a user logs into the client system. If there is an update pending, the server initiates a "pull," downloads the update, installs the update, and then restarts the application, if necessary. In 2010, DPAS negotiated to include the DPAS software client on the defense gold disc for installation on all defense seats, but this required Legacy Application Deployment Readiness Activity (LADRA) testing and was phased out within a few years.

It is widespread practice in the software industry to move away from client/server-based systems and toward web-based applications. Client/server-based systems require installation on each asset that a customer wishes to use the application on. This can cause access issues if there are problems with specific assets and customers need to use "loaner" assets. As discussed earlier, clients require updates which causes a delay in use of the system whenever an update gets posted to the server as clients must download, install, and restart the application. All processing is carried out on the client's individual asset and speed is related to the asset hardware configuration. Finally, servers must be taken offline to install updates, which requires late night or weekend work from the server management team. Web-based applications do not require updates to be pushed to a client as these types of applications are accessed directly through a web-browser, such as Google Chrome or Microsoft Edge. Access via a web-browser also removes the installation barrier that client/server-based applications face. Updates can be made swiftly to the application, many times without any downtime accrued by the server. Web-based application servers deliver dynamic content in much the same way that the client/server model operates with the benefit of a high control of hardware performance, as the performance of the system is not restricted by the asset that the customer accesses the system from. Web-server applications come with the added benefit of access control and dynamic presentation built into the system, allowing sites to restrict access only to authorized users and to present only the data that a customer is permitted to view. With these points in mind, it is clear why the DPAS stakeholders desired to move to a web-based application versus the client/server-based architecture.

6.3.2 DPAS 2.0–3.0 Transition

In 2017, it was decided that to overcome the challenges presented with a client-based system, the DPAS system would transition to a web-based system. The development team proceeded to work on a prototype, transitioning the client/server-based DPAS system to a sustainment only state. The proposal to transition as presented to the prime stakeholders of the system and approval to develop requirements was given in 2019, though prototype development began as early as 2016.

Managing a system upgrade while the current system is in place requires a heavy effort on software developers. Not only does new architecture have to be designed and deployed, but the previous obsolete client must still be supported. This caused the development team to split into two sub-teams. One senior developer was allocated to maintain the current client-based system. The remaining senior developer, journeyman developer, and new college graduate took on the effort to design the web-based system. Other team members included a Database Administrator, the Project Manager, a Customer Service Support Specialist, and a Software Tester.

The team decided to transition to an Agile Scrum project execution. This was due to the team attending an Agile Scrum training session and seeing the benefits that Scrum can bring to a development effort. The team stood up the initial Scrum Team with the Software Tester taking on the role of Scrum Master. The software developers joined the team. It is important to note that there was not a Product Owner designated, though the Project Manager would have been the natural selection for this role. Having a Product Owner is essential to the Scrum process. This is a crucial point to understand as it is one of the key drivers of the issues that the team was soon to experience.

To transition to Agile Scrum, the team had to create a repository for a Product Backlog, as is necessary for Agile Scrum development. The team chose to use GitLab for their backlog management. In GitLab, teams can create and use dynamically generated issue lists, populated with work items. Epics were created as GitLab epic work item types and User Stories were created as GitLab tasks. Sprints were set up as milestones in GitLab, as is standard practice. The Scrum Master owned and managed the backlog.

After populating the project backlog, the team began planning and executing Sprints. The backlog was prioritized according to the team's view of customer needs. This differentiates from pure Agile Scrum, as the team should have had a Product Owner that prioritized the backlog. Instead, the team used their own opinion of customer needs and desires, as well as developer needs and desires, and prioritized the backlog accordingly. At each Sprint Planning, the team took whatever work items were on the top of the prioritized backlog, counted down the backlog until they reached what could be done within their velocity, and added those items to the Sprint Backlog. By restricting themselves fully to a 100% prioritized backlog, the team lowered their capability to execute what was needed in development, instead focusing on the artificial prioritization represented in the backlog.

During the execution of the Sprints, work items were claimed by developers and moved across the Scrum Board, being marked as "In Progress," "In Review," and "Closed." A Scrum Board is an Agile Scrum project management tool that allows Scrum Teams to have a visual representation of the planned work for a Sprint to increase communication and transparency during project execution. As the team was focused on performing in a self-managing method, this allowed the team to track current work item status and have transparency to the project stakeholders. The board also allowed the team to increase collaboration through ticket comments, tagging, and other matters related to the work. Testing and code review was performed as work items were populated into the "In Review" column. Code review required another developer to review the code related to the ticket item for potential bugs, "cleanliness," and adherence to policy standards. Testing activities were performed by a dedicated tester.

6.3.3 The Problem with Too Much Agility

From 2016 to 2017, the DPAS team made little headway toward project completion. During this period, the team was able to only accomplish the creation of a login page and a landing page. The Project Lead and Branch Manager were concerned with the progress of the project as most of the development team had been allocated to the development of the new iteration of DPAS. An analysis of the cause of the deficient performance by the team is detailed below, but it should be noted that these issues are not unique to this team and are a common occurrence in software development projects, especially in teams that are new to Agile Scrum.

For Agile Scrum to be successful, the Scrum Team must have the dedication to follow the Scrum process rigorously and through a standardized methodology. Agile Scrum is the industry standard as it has had decades of adjustment to meet changing customer needs and has a proven record of success when applied thoroughly. This is not to say that all Artifacts and Events meet every single goal of the Scrum Team, but without a thoughtful and reasonable assessment of the Scrum methodology with regards to a specific project, every effort to deviate from the norm should be avoided. Implementing Agile Scrum does not require the abandonment of long-term planning, documentation, and structured project management. Rather, a culture of embracing change in a reasonable and thoughtful manner should become part of the corporate culture.

The DPAS team took some of the tenets of the Agile Manifesto to heart and, in many ways, quite literally while ignoring others. After their initial Scrum training, the team had a large poster of the Agile Manifesto printed out and hung on the wall of their cubical space. This was to reinforce the idea that being Agile was of utmost importance. Unfortunately, the team did not have a member onboard who had experience in the successful implementation of Scrum and therefore were novices in their adaptation and execution. Certain team members used this time to assert their ability to work, however they wished and disregarded any form of engineering and technical rigor. It was this last part that led to the lack of deliveries and project execution.

By focusing too much on the letter of the law, specifically with regards to the Agile value of "Responding to change over following a plan," the team entered a cycle of constant change without any successful deliveries. The team was in a constant state of change due to the literal interpretation of this statement. If a team member found a new coding standard or methodology, even a new coding language, they would refactor all the code base to work in the new way. This resulted in the DPAS code being refactored multiple times. The team decided to convert from using JavaScript to a more succinct version called CoffeeScript, which utilizes a structure more akin to that present in Python coding rather than being presented as a C derivative. By implementing this new language in the middle of the development effort, all the developers had to spend time not only refactoring the entire software solution but also learning the new scripting language conventions. This had a direct and negative impact on the project's cost, schedule, and performance.

This was not the only occurrence of too much change in the project execution but is notable as it reflects the norm in execution.

Another example of leaning too much into the Agile Manifesto in a literal sense is the team's approach to "Working software over comprehensive documentation." The team took the approach of adding no internal software documentation, such as JavaDoc or code comments, under the assumption that any code written should be done in such a way as to be self-documenting. The core idea of self-documenting code is that no internal description of the functionality present in a method is required as the variables and structure are written in such a way that the intention of the author is revealed through descriptive names and careful structuring. The team disregarded common self-documenting guidelines, such as external documentation as defined as a header above a class definition that describes various internal components to the class, such as the title, functional description, coding author, date modified, etc. Comments are still needed in self-documenting code, though not necessarily as verbose as presented in a non-self-documenting way. Example topics of comments include the need to describe organization, insights into complex methods and algorithms, assumptions or constraints, and examples of use, when feasible and reasonable. The team disregarded all these factors and instead commented and documented nothing in the code, resulting in a code base that had no historical documentation of developer thoughts or explanations, limiting the ability of other developers to gain swift understanding of the function of the code being reviewed.

"Customer collaboration over contract negotiations" is one of the core values listed in the Agile Manifesto and therefore a core component of Scrum which the DPAS team did not adhere to. Agile Scrum focuses on creating an environment where stakeholders and developers work closely together to create a working piece of software in an efficient manner while maintaining the ability to pivot to changing needs. The Product Owner holds the critical role of representing and facilitating customer interaction for a Scrum Team. As stated earlier, the DPAS team did not have a Product Owner and therefore had little to no representation of customer needs during critical Scrum events. This is especially relevant when reflecting upon the Product Backlog and Sprint Backlogs. Without a Product Owner prioritizing the Product Backlog with customer needs, the backlogs are left to be prioritized by the development team and their understanding or assumptions of what the customer desires most and without the benefit of the customer collaboration feedback loop. This leads to a situation where developers are performing software development and major product decisions with little true insight into customer needs, goals, requirements, or priorities. Without direct customer interaction, developers and engineers are left to guess what a customer needs, often resulting in a less-than-ideal end-state.

Finally, the team misunderstood "Individuals and interactions over processes and tools." The DPAS team focused on the desires of the individuals internal to the team and not communication and interaction with those outside of the team. They also put an undue emphasis on the tools and Agile process that they were using. This removed the focus from the delivery of a high quality and working product that would meet customer needs. Instead, the solution was treated like an experimental sandbox with frequently changing tools and processes. This directly led to the equivalent of negative velocity. To clarify, the team appeared to be making forward progress as they were creating and testing code, but the code that was created and tested was often rewritten solutions, implementing innovative ideas, concepts, and coding standards, resulting in almost constant rework. Rework such as this is difficult to track when solely looking at velocity. One must understand that the code being delivered to the integration servers was reworked and not representative of positive momentum toward final product delivery.

As stated before, the DPAS team's failures to adopt Agile Scrum in a productive manner is not uncommon with development teams. Surveyed customers report projects developed under an Agile umbrella as unsuccessful roughly 50% of the time within the boundaries of cost, schedule, and performance [5]. When transitioning from a traditional or predictive methodology, such as Waterfall, it is important to plan out the process transition and stick to the plan. It is equally important to have someone attached to the project to "steer the rudder," keeping the project on track for delivery, normally represented by the Product Owner. Finally, in Agile Scrum, it is essential to have a cooperative social environment where communication and collaboration are put to the forefront through bi-directional trust. Development team members cannot work independently and must instead work closely with the rest of the team, checking in frequently and sharing ideas. When a team does not execute collaboratively working toward a unified product vision, Agile projects fail.

6.3.4 Results of Implementing Systems Engineering Rigor in the DPAS System

In 2019, it was determined that the DPAS team needed more structure. A new project manager was put in place that took on the role of Product Owner. The project manager reviewed the Agile Scrum standards and created a solidified project vision. The team was required to communicate with leadership through the Product Owner to ensure that development standards were met and ensure that system development reflected customer needs. Changes to software execution and coding standards were limited only to those that were required to facilitate delivery of the software. This led to the DPAS 3.0 project making forward momentum toward delivery.

One of the first steps taken by the team was to work collaboratively with the customer to generate a Software Requirements Document (SRD) containing high-level business requirements for the DPAS 3.0 project. The requirements were generated with customer collaboration and concurrence. A formal SRD was created and presented to the project stakeholders with each stakeholder indicating approval of the requirements through a signature. The customers represented a broad range of facilities, including Intermediate Maintenance Activities, In-Service Engineering

Activities, the Program Manager, and the Project Manager. While the requirements are only requirements through a broad definition of the term requirement, those listed in the SRD did solidify the project goals and needs of the customer. The SRD also gave the Scrum Team a starting point for conversations, allowing for ease of clarification and further requirement discovery.

By creating and executing software development in a more structured Agile Scrum method, the team was able to work together more effectively. Previously, the team embraced far too much agility, taking the definition of agility to mean doing whatever they wanted, whenever they wanted to. While having the ability to quickly respond to changing customer needs and requirements is necessary for an Agile Scrum project to be successful, a lack of any form of structure almost certainly leads to the failure of a project. The lack of a form of structure and instead promoting a complete focus on fluidity in changing methods can be defined as an anti-pattern in Agile when considering an anti-pattern to be defined as a practice that could initially be thought of an Agile, but proves to be ineffective in execution, often having drawbacks that outweigh any perceived benefits.

Close team collaboration is essential to the success of Agile Scrum. Moreover, historically, the lack of unified team alignment and a focus on individual software development have been identified as a cause of failures in Agile projects. The lack of close collaboration is readily apparent in the DPAS project. Team members would take Scrum work items on their own and then independently develop them. Pair programming was essentially non-existent. Senior developers actively avoided sitting with junior developers, leading to a lack of essential knowledge transfer and the ability to quickly overcome challenging programming problems. Another consideration of pair programming is that while pairing, developers do not necessarily output more lines of code than individually working, the code is of a higher quality and less defects are introduced into the general production system as by having a second set of eyes watching the code being developed, there is a built in quality check that would otherwise be absent. While it might not be readily apparent that close collaboration with customers is a systems engineering focus, it is one of the key tenets of the systems engineering discipline. The goal of systems engineering is to create a holistic system that meets customer needs while bringing together cross-functional and multi-disciplinary teams. By reducing the instances of developers working in individual silos, the DPAS team was able to better develop code and deliver a higher quality product.

Finally, the project manager implemented frequent customer check-ins. These check ins took the form of "demos" where the development team would present modules and subsystems that were close to ready to be deployed. This allowed customers to give feedback prior to publish to production. By demonstrating the working software to the customer prior to release, new requirements were generated, and flaws were identified with time to fix any issues prior to final publication. This has led to a closer and more successful collaborative environment with the DPAS team and their customers.

The path from DPAS 2.0 to DPAS 3.0 has been a difficult one for the DPAS team, but through the embracing of structured Agile Scrum and systems engineering methodologies, such as requirements management, the team has overcome the previous conflicts and struggles and has a clear path forward to final publication. This required a firm hand on the rudder to steer the team to success and ensure compliance with policy, process, and standards. The team had to learn to work together instead of focusing on individual contributions. An understanding that the perfect solution is not the goal, but rather a solution that makes the customers happy and can be delivered in a timely manner, was adopted by the DPAS development team. Through the implementation of these changes and at the time of writing, the team has drawn closer than ever to the final publication of DPAS 3.0 and closing out this almost 10-year development effort.

6.4 Case Study 4: How Too Much Rigor Can Stifle Innovation as Shown in a Defense Qualification Management System (DQMS)

6.4.1 Introduction

The DQMS application is a web-based training-support software application containing code, datastores, and tables that are utilized to track technical training throughout servicemembers' careers. The system is used to validate knowledge and skill sets through short- and long-term individual or group training plans, qualification tracking, and testing. DQMS provides a feedback system so qualification supervisors can provide feedback to qualification candidates. The hosting of the system is done through the utilization of COTS hardware and support software. DQMS currently supports more than 250,000 user accounts, both military and civilian. The system provides tracking for over three million training tasks and qualifications, certifications, and licenses (QCL).

6.4.2 Approach to Systems and Software Engineering

The DQMS project is managed through a very rigorous, detailed, and documented system and software engineering approach. As is required for DoD Acquisition programs, the process is laid out in a Systems Engineering Plan. The plan requires the system to follow a tailored Defense Unique Software Intensive Model. Six Systems Engineering Technical Reviews are held for each planned software release. The team utilizes a modified Agile Scrum methodology. Customer interactions are through project leadership and formally held during Integrated Product Team (IPT) meetings. DQMS system configurations are tracked through Product Baselines. Each version is planned out, presented, and approved by the Program Office through a Software Requirements Specification (SRS) which is revised at a Software

Requirements Review (SRR). The project utilizes a very rigorous systems engineering methodology, which is unusual for most software projects.

The DQMS Systems Engineering Plan (SEP)[3] states that the DQMS application utilizes a tailored Adaptive Acquisition Framework (AAF)[4] Defense Unique Intensive Model modified to support a project in a sustainment status. The model used is not actually from the Adaptive Acquisition Framework Software Acquisition pathway, but rather a modification of the Major Capability Acquisition. This is most likely due to the author of the SEP not fully understanding the changes presented in the AAF update and rather utilizing the older Software Intensive pathway. This acquisition pathway requires a large amount of technical and engineering rigor per DoD requirements as proscribed by the Office of the Undersecretary of Defense for Acquisition and Sustainment and detailed in the "Operation of the Defense Acquisition System", as shown in Fig. 6.7 [6]. Milestone Decision and Systems Engineering Technical Reviews (SETR)[5] are held before any publication can occur. Versions are fully planned prior to implementation. The DQMS project performs six SETR events during each version execution cycle, a Software Requirements Review (SRR), Critical Design Review (CDR), Functional Configuration Audit (FCA),

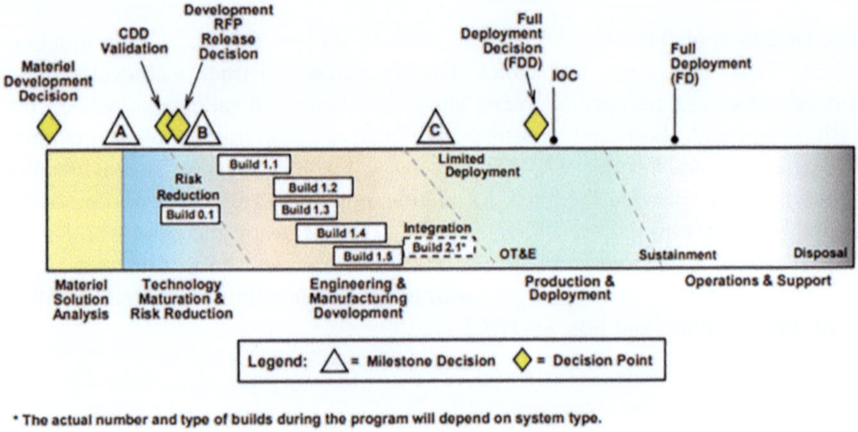

Fig. 6.7 Defense unique software intensive program

[3] A Systems Engineering Plan (SEP) is a living document that provides the blueprint for the technical aspects of a program, detailing the strategy for conducting, managing, and controlling systems engineering activities from concept to disposal.

[4] The Adaptive Acquisition Framework (AAF) is the U.S. Department of Defense (DoD) system for acquiring high value goods and services, replacing the traditional "one-size-fits-all" of the previous Acquisition Framework with tailored acquisition pathways based on a program's unique characteristics.

[5] Systems Engineering Technical Reviews (SETR) are events held for defense systems where senior technical leadership evaluates the current state of a system in development to determine whether adequate engineering and technical rigor are being met, per U.S. defense policy.

Physical Configuration Audit (PCA), Production Readiness Review (PRR), and Project Performance Reviews (PPR) in place of the standard Release Backlog Reviews (RBR).

At the PPRs, risk assessments are performed as well as communicating Earned Value Management (EVM) metrics and status and Size/Stability, Cost/Schedule, Quality, and Organization metrics. These metrics are important, as breaks in thresholds for these metrics are triggers to take action to update or change the project and reflect a very rigorous attempt to keep the DQMS project within cost, schedule, and performance thresholds, as given in Table 6.3. Of note is the Organizational metric, which has a threshold of "Any deviation from the planned burndown." This requires an explanation to the stakeholders as to why the Predictively planned burndown has been deviated from and how the development team will recover to stay on track for scheduled publication.

The DQMS SEP details a procedure for documenting and controlling software enhancement requirements. These requirements are generated at IPTs and documented in an SRS. The requirements are reviewed and accepted by a stakeholder team consisting of a customer, Technical Project Manager, Customer Advocate, Configuration Management Manager, and Systems Engineer during an SRR. If a program office is involved or affected, the specified Program Manager must also sign the SRS. The signatures in the SRS represent acceptance of the requirements. These are the requirements that the software development team will use to perform changes to the DQMS system.

Specific design reviews are required by the DQMS SEP. A Preliminary Design Review is an "independent assessment that informs the Milestone Decision Authority (MDA) of the technical risks and the program's readiness to proceed into detailed design" [7]. According to the DoD "Systems Engineering Guidebook," this review is used to ensure that the initial design and architecture are adequate to meet the requirements of the stakeholder and that the Milestone Decision Authority believes that the system can be instantiated within the constraints of cost and schedule goals [8]. A Critical Design Review is "an independent assessment that informs the MDA of the technical risks and the program's readiness to proceed into fabrication, system integration, demonstration and test" [7]. The goal of the CDR is to ensure that the system is stable, operates within proscribed requirements, and meets mission goals in preparation for publication to the warfighter and provides

Table 6.3 DQMS software programmatic performance metrics

Technical performance measure	Objective	Threshold
Stability/sizing—Requirements variability	Zero variability	<10% of planned requirement increase
Cost/schedule—PPR EVM	1.0	Lower threshold = 0.9
Upper threshold = 1.1		
Quality—defect open vs closed	New defects = closed defects	When defects created out, number defects closed
Organizational—full project Sprint burn down	1.0	Any deviation from the planned burn down

stakeholders the confidence needed to proceed with final delivery [8]. These reviews are event driven and occur a single time in the lifecycle of the system enhancement project. They are technical reviews used to grant permission for continuance of a project.

6.4.3 *The Problem with Too Much Rigor*

The DQMS SEP states that the system is designed and developed using a modified AAF framework. This is untrue, as the AAF Software Acquisition Framework does not utilize the same pathway as laid out in the Major Capability Acquisition Pathway. It appears that the DQMS Systems Engineer did not update the SEP to reflect the changes that make the AAF "adaptive" rather than very static and predictive, as the old Defense Acquisition System operated. This approach utilizes a Predictive methodology, requiring formal customer approval for any changes to deliverables. Planned versions are large efforts and the system is unable to pivot for minor changes. There is little capability for the project to use iterative deployments or adapt to changes in scope and requirements. As can be clearly seen in Fig. 6.8 [3, 9], the current AAF Software Acquisition methodology focuses on iterative development with active user engagement [3, 9]. While design architectures are mandated, it is done in an incremental and iterative manner, allowing for rapid changes due to customer needs, goals, and requirements being updated through stakeholder communication and engagement. As detailed in the DoD Instruction 5000.2, "Operation of the Adaptive Acquisition Framework", when planned and executed in a Predictive

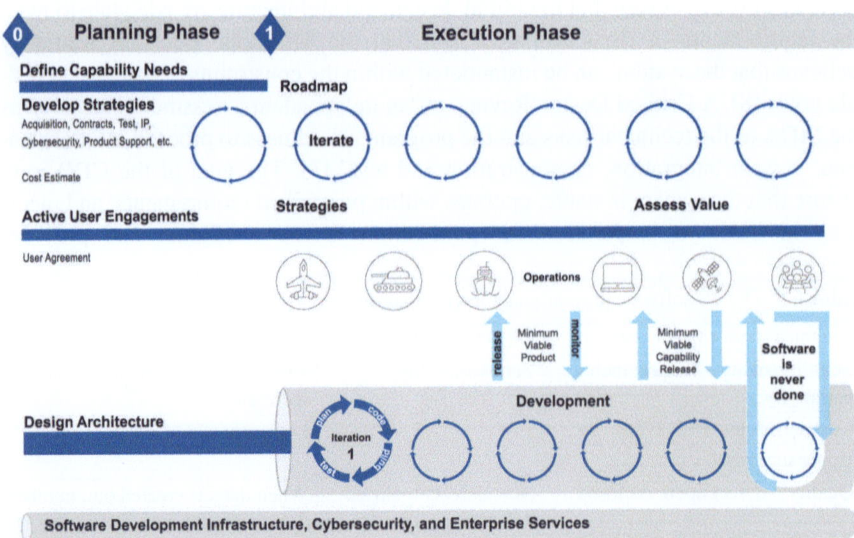

Fig. 6.8 Adaptive acquisition framework software acquisition pathway

manner, such as presented in Fig. 6.8, this capability is lost, as any changes break the model.

The approach the DQMS project team takes to track project performance and health is very predictive and onerous. It is difficult to track project performance utilizing metrics such as EVM, as EVM requires the use of the Agile anti-pattern "Big Modeling Up Front" to create a comprehensive model of the requirements of a system, analysis of the requirements, an architecture to execute the requirements to, and a detailed design prior to executing the development of the system. This is not an Agile or iterative approach, but rather a Predictive approach to design which has been proven to be ineffective for software system development. In many projects, software developed under a Predictive, or Waterfall, methodology either end up not being deployed or fall well outside the acceptable parameters of cost, schedule, and performance due to the inability to swiftly pivot and adapt to changing customer needs. The larger the system, the harder it is to fully capture system complexity in requirements documents and models, as well as stakeholder understanding of what they need versus what they think they need. Adding to this is a very tight tracking of deviations of a fully upfront planned Project Sprint Burndown. This assumes that all work is known prior to the execution of development and is another anti-pattern in Agile, allowing for no flexibility in delivering early or late, depending on changes to customer needs or desires. As is clear, the DQMS project suffers from too much rigidity in project execution.

The way that the DQMS project team manages requirements through a formal signature process in an SRS can be problematic in software development. Software requirements need to be fluid and adaptable to allow for changes in customer needs and the evolution of the system concept as development progresses. This is due to the complex nature of software engineering. Software engineering is as much an art as it is an engineering effort. Requirements management must be flexible and adaptable when considering that many customers do not know what they need until they see it or change their mind as new material is presented to them. Utilizing systems such as JIRA to document, track, approve, and maintain status of requirements is much more aligned with the Agile principles than the more Predictive methodology of hard documentation through a formal signature process. In the process that the DQMS project team uses, each time a customer wants a new change request, or a change to an existing change request, the stakeholder team must fully meet, deliberate, create a new SRS, judge the validity of the SRS, and hold a formal signature meeting. This is not only highly inefficient, but also very costly to the government as multiple work hours are exhausted in this effort, usually with prominent level, and highly paid, government employees requiring attendance.

Formalized project execution documentation, such as SRSs, can lead to the stifling of communication between a development team and the customers. Agile projects focus highly on customer and developer collaboration. One of the key principles of the Agile Manifesto is "Customer Collaboration over Contract Negotiation." The reason behind this is that the goal of any software project is to deliver what the customer needs when the customer needs it, not what the business team negotiates to build, regardless of customer need. Communication between individuals, be it stakeholders, customers, business managers, or developers, is an evolutionary act and has many obstacles that can stand in the way of full understanding. Communication needs cover many facets of a project, such as goals, objectives, tasks, procedures, constraints, interdependencies, requirements, timelines, prioritization, responsibilities, accountability, models, and even the final system solution. It is seemingly impossible to assess and document the "perfect" system when accounting for all these variables. Therefore, communication through collaboration is required leading to an evolutionary development of a system, rather than a static one. This not only creates a better understanding of the system context, but also builds trust between collaboration partners, through the reinforcement that evolves from consistent and continuous feedback loops.

The DQMS SEP requires formalized technical reviews with specified input and output criteria in the form of the Preliminary Design Review (PDR) and Critical Design Review (CDR). This is against current DoD Software Acquisition Pathway intentions. Waiting to demonstrate a software product to a customer until it is ready for a technical review style meeting is counterproductive and against the collaborative nature of Agile development. DoD Instruction 5000.87, the "Operation of the Software Acquisition Pathway," proscribes close collaboration with stakeholders using frequent product demonstrations, leading to the capability of continuous delivery of capabilities through incremental development [10]. The use of two technical reviews stymies the collaborative capabilities of the project team and does not take advantage of the ability for software to be delivered incrementally. In this instance, smaller releases containing individual change requests can be reviewed and published individually through smaller review boards rather than waiting for large, project sized deliveries.

6.4.4 DQMS Case Study Conclusions

The DQMS system suffers from adherence to a too rigorous engineering methodology. The systems engineering principles that are laid out in the DQMS SEP align more closely to what one would expect in a hardware centric system, rather than a software development effort. The standards taken to meet DoD Acquisition standards are antiquated and do not reflect the updated Adaptive Acquisition Framework policy. Much of the policy laid out in the DQMS SEP requires formalized project documentation and a high amount of technical rigor that is more akin to a Predictive project. While the DQMS SEP does reference daily Scrum meetings, there is no other reference to Scrum or Agile, nor does it lay out management of User Stories, requirements elicitation, or Story Points. The DQMS project suffers from a too Predictive project execution style and would benefit greatly from the adoption of Agile practices, specifically Agile Scrum, utilizing requirements traced to User Stories, closer customer collaboration, and increased developer independence.

The DQMS project team is working on employing more Agile techniques to enable more efficient and flexible development. They are planning to release new versions monthly to ensure rapid deployment to the fleet and enable enhanced communication and feedback from customers. The team is stopping fully planning releases and instead implementing a system closer aligned to Continuous Improvement / Continuous Development (CI/CD). This brings the project closer to alignment with the principles laid out by the DoD Chief Information Officer in the "DoD Enterprise DevSecOps Strategy Guide" [11]. The Product Backlog items, specifically defects, have currently all been prioritized, and the team is working on the defects in priority order, bringing the Backlog into compliance with Agile Scrum ideals. As is readily apparent, the team is working diligently to transition to a more Agile focus that can swiftly adapt and react to changing customer needs.

6.5 Case Study 5: Adapting a Work Management System to Modern Software and Systems Engineering Practices

6.5.1 Introduction

The DoD Work Management System (DWMS) is a web-based application initially developed in 1998 as a workload management system designed to manage projects and processes within Department of Defense workshops. The initial system was an Access-based application for internal use in the DoD. In 2002, the DWMS application transitioned to a web-based client server application. In 2004, DWMS expanded to allow for use by external customers. It has since gone through two more versions. The DWMS development team is currently working on an updated version but has struggled to formulate a suitable and acceptable solution for this fourth version of the software.

6.5.2 DWMS Software Development Process

The DWMS team has done a respectable job of staying up to date on modern software development techniques, implementing industry standards in a thoughtful and impactful manner. The team uses the Agile Scrum framework to develop their software projects. The project has documented standards explaining the implementation of Agile Scrum that is used, as well as workflow expectations and project roles and responsibilities. Quality controls are laid out and defined as well as testing procedures. The DWMS team utilizes a work tracking system, JIRA, to plan and execute Sprints. Various checklists are used as gates to ensure compliance with DWMS development standards. Overall, the documentation is adequate for the DWMS team.

The two core documents used to set policy for software and systems engineering for the DWMS project are the Configuration Management Plan and the DWMS

Feature Development Process Standard Operating Procedures. The Configuration Management Plan lays out how the development team sets the standard in which configuration will be controlled, including specified meetings, change requests, requirements traceability, definition of configuration items, test procedure documentation, and software release procedures. The DWMS Feature Development Process Standard Operating Procedures detail DWMS process roles, Sprint definitions, Sprint workflow states, the feature development process, the JIRA management process, the GIT management process, and has appendixes with various guidelines. These two documents set a high-level standard for development efforts. The areas covered are adequate to meet high-level expectations.

The team does a better than average job of knowledge management with the project. There is a repository with documents, operating procedures, and guidance for new employees as well as standard operating procedures for setting up Integrated Development Environments (IDE) and connecting to servers. In depth documentation is present for the deployment process, test case templates, and various miscellaneous information such as technical API guides, user guides, and other data.

The DWMS team has various roles associated with the project, many of which align to traditional Agile Scrum roles. The Code Owner role is defined as the technical leader and is responsible for providing technical direction and vision, code implementation oversight, and setting coding standards. Developers are responsible for formulating software solutions for development efforts, creating and managing test cases, managing merge requests in GIT, and maintaining development item status in Jira. The Project Manager is aligned with the Product Owner with the responsibility of determining features and priorities, setting schedules, and acting as a liaison between the customer and the development team. The Scrum Master is responsible for helping the Project Manager and Code Owner set up and maintain the current Sprint; this role can be filled by any team member. Customer Testers are customers, or stakeholders, who are available to the development team to perform requirement and functional testing for features they are SMEs for. A Deployer is a person responsible for packaging application updates for deployment as well as updating SQL schemas and objects. Reviewers are listed as team members who perform code reviews for other team members prior to a code check in. Information Assurance (IA) personnel ensure that updates are made to application servers as well as ensuring cybersecurity requirements are met. Finally, Change Review Board (CRB) members are customers representatives who assist in steering the direction of software development for product evolution. As is clear, there are quite a few roles involved in the development of the DWMS product, which feeds into the complexity of customer and stakeholder communication as well as the complexity of the development environment with consideration of hierarchical responsibilities.

The DWMS team utilizes Agile Scrum in a modified fashion. They ascribe to a normalized Sprint schedule, with planned Sprint ceremonies, or events. The Sprint ceremonies performed by the team are the Sprint Retrospective, Sprint Planning, Scrums, which are referred to in documentation as Stand Ups, Sprint Reviews, and the Backlog Grooming. The development team utilizes Story Points to gauge the size of a task, as given in Table 6.4. Of note, sizing is directly correlated to days to

Table 6.4 DWMS story point schema

Size	Days	Points
Small	1–2	1
Medium	3–5	2
Large	6–8	3
X-large	9–21	4
Epic	>21	5

execute, which is different from normal Story Pointing, as the normal procedure is basing Story Points on complexity, not time of effort. The Jira Sprint Workflow is overly complex, with many steps required prior to deployment. Of note, the Code Owner must review all work prior to submission for final approval.

Configuration items are detailed in the Configuration Management plan as well as the expectations for configuration management. The Configuration Management plan details the policy for version control. Change requests and defect identification are noted as being generated by users. Test procedures are detailed, along with the management of test cases and where they are stored. Requirements for the creation of design documents, as well as traceability from the documents to functionality, are discussed.

DWMS uses dirty trunk, clean branch development version control methodology. In this method, the development team pushes all code to a centralized trunk. This code is then baselined and segmented off into a branch which is then used for publishing to a production environment. The main branch represents the most recent version tag along with recent feature and defect solutions since the last baseline. All items that are successfully tested and approved are pushed to the main branch. These changes are bundled, deployed, and then tested in production. After successfully passing a test event in production, the system version is updated, and the main branch is baselined.

Testing is touched upon both in the Configuration Management document as well as in the DWMS Feature Development Process Standard Operating Procedures. The testing policy does not have a formal Test Engineer identified. Rather, the development team performs testing. The procedures state that the originating developer will test their solution themselves and then pass the solution over to another developer for testing in the test environment. User testing comes next and is verified by the project lead. Unit tests are built into the system and are required Sprint Tasks for each Sprint Item. Most of the regression testing is performed through unit tests. Unit testing is the process of performing automated and coded tests during the process of the code to the repository, which is often utilized in CI/CD pipelines.

Requirements management is not discussed in the DWMS documentation, outside of reference to design documents. The documents do refer to requirements when discussing testing, in stating that tests verify requirements were implemented and working correctly, but there is no reference to what form the requirements take. User Stories are also referenced but no documentation or policy surrounding the creation, verbiage, or tracking of User Stories. The documentation is written with the appearance of a baseline understanding of what requirements and User Stories are and makes no effort to define the structure or use of either type of statement.

6.5.3 Difficulties with Partially Implemented Agile Scrum and Engineering Rigor

The DWMS project suffers from partial implementation of the Agile Scrum Framework and engineering rigor. While many recommended processes and policies are in place, they are written at a very high-level resulting in little detail being communicated. Many of the policies make assumptions of the readers' knowledge and understanding. When viewed as an impartial external entity assumed to have no understanding of a software development lifecycle, structure, or Agile Scrum knowledge, the policies do not function well.

The first problematic part of the policies is found within the roles section of the DWMS Feature Development Process Standard Operating Procedures. Many roles are detailed and are not part of the Agile Scrum framework. Certain roles, such as the Product Owner, are misaligned. The Project Manager can, and often does, serve as the Product Owner. However, the Product Owner should not be responsible for the creation or management of Sprints. That is the job of the Scrum Master and Scrum Team. The Code Owner role is especially problematic as this is not a role proscribed by Agile Scrum. It appears to be a control role that fills duties that are Product Owner duties, such as providing technical direction and vision. Due to the requirement that all checked in code be reviewed by the Code Owner, it also causes an unnecessary bottleneck in the development cycle. The Team Tester role can be filled by developers, which does not leverage the strengths of having a full-time Test Engineer on staff, but rather uses Developers who are not formally trained in the role. The Scrum Master role is not staffed as an individual but rather can be filled by any team member. This is not ideal and can lead to serious issues as the Scrum Master is supposed to be the evangelist and the authority on Agile Scrum execution. The Scrum Master is accountable for the Scrum Team's effectiveness and for establishing Scrum as defined in the Scrum Guide. If no training or experience with Agile Scrum is required, then the role becomes inconsequential. The roles need to be streamlined, brought into alignment with the Scrum Guide, and the Code Owner role should be removed entirely with the duties given to the Product Owner.

Another area that needs improvement is the way the team uses Story Points. Story Points are units of measure designed to estimate effort as described relative to work complexity, the amount of work, and the risk involved in the work. The DWMS team has linked Story Point estimation directly to the number of hours required to complete a task. This does not work well in a team of mixed-experience engineers. For a junior engineer to perform a task, it might require 4 h. A senior engineer with experience in fixing similar issues may be able to complete the task within an hour or less. In this case, the estimation falls apart as each engineer will estimate the work at a different level. Expanding this out to a project estimation makes the task impossible to estimate as complexity is not considered nor is risk considered. Rather than basing Story Points on a set time to perform work, the team should transition to the industry standard of basing Story Points on complexity, amount of work to complete the task, and risk involved.

With regards to the Agile Scrum policy, the DWMS team is mostly following guidance as laid out in the Scrum Guide. One major differentiation is the reference to the Daily Scrum as a Daily Standup. The Daily Scrum is specifically not a stand up, as the Daily Scrum has rules laid out in the Scrum Guide, such as a restriction to 15 minutes, who participates, and what is discussed by the team. A stand-up can be anything used to communicate within a team and can include management and other individuals who are not formally part of the Scrum Team. It is not a status meeting for the Scrum Master; it is a synchronization meeting for the Scrum Team. Beyond changing the name of the meeting to Daily Scrum and adhering to the Scrum Guide rules to manage the meeting, it is common for Scrum Teams to use the "three questions" to focus the discussion had at the Daily Scrum. The three questions are "What did I work on yesterday? What am I working on today? and What issues are blocking me?" By focusing on what the individual did yesterday, what they plan on doing today, and what is standing in their way, all the team understands the progress the team has made toward completion of the Sprint Goal and any blockers that might be problematic for the team. This is especially helpful as it gives the Scrum Masters notice that there is a blocking issue that needs solved. Clarifying expectations at the Daily Scrum and following the guidance laid out in the Scrum Guide should result in increased efficiency and communication within the development team.

The single largest concern with the DWMS software implementation planning is the lack of formal requirements management documentation. For the DWMS program, requirements are listed in a Software Requirements Document. The requirements are not written in a standardized language format, such as INCOSE style shall statements. Neither are the requirements following traditional Connextra User Story format. Requirements are not tracked in a system that is query-able, nor are they clearly demarcated through a numbering schema. As written, the requirements do not meet even the most basic requirement writing guidelines proscribed by INCOSE, including testability. The requirements also include extra elements, such as data elements, written in abbreviations with no clear definition of what the elements are. An example of this is a data element named "Prob Factor." Prob Factor is associated with an element called "Quick Quotes." There is no explanation of the form of probability that should be shown, the customer driven use or rationale, nor any way to rationally write a test case to ensure the requirement is being met by a system implementation. This is the standard, not the exception, to how the requirements for DWMS are written. A full system "as built" requirements documentation effort needs to be undertaken to build the foundation for future systems engineering work, otherwise the system will continue to fail to meet customer needs, goals, and requirements.

6.5.4 DWMS Case Study Conclusions

The DWMS project has good practices built into the team's processes and policy, as well as poor practices. It is obvious that the DWMS team has worked hard to implement engineering rigor in the best way they know how. They have made efforts to document requirements in a method that is not too far from standard in software engineering. The team has implemented a form of Agile Scrum, though not as fully implemented as it should be. This is probably due to no one on the team being certified as a Scrum Master or Product Owner. With guidance and work, the DWMS project can swiftly be brought into alignment with systems engineering focused Agile Scrum development.

As of the time of writing this case study, the DWMS team is making efforts to implement formalized requirements management. The team has been trained by their Principal Engineer, a formally trained systems engineer, in INCOSE style requirements management. The team is focusing efforts on creating an in-depth requirements document for the updated version of the DWMS project. As the original DWMS program is being sunset once the updated version of DWMS is implemented, they are using the time to focus on the next version, rather than documenting functionality that will not be ported over. When the team has this requirements document and a requirements management plan in place, they will have built the foundation upon which to implement increased engineering rigor for the future and more thoroughly understand customer needs, goals, and requirements.

6.6 Case Study 6: Successes in Designing an Enterprise Logistics Software System (ELSS) Utilizing the SEFAD Process

6.6.1 Introduction

In the early 2020s, the DoD set about creating an all-encompassing enterprise logistics system. This goal of the system was to absorb the functionalities of all of the other existing logistics systems across the agency. The reasoning behind this was to not only attain cost savings through the de-duplication of work efforts and sustainment of multiple systems performing the same core functions, but also to leverage the weight of the whole of the DoD for contract negotiations. When one large agency negotiates data, it is much more cost efficient than each individual working-level agency performing the same negotiation, thereby realizing cost savings for the taxpayer. The ELSS effort was not unprecedented, but it did come at a time of frequent large scale enterprise government systems rolling out in a failed method.

6.6.2 Program Instantiation

The ELSS program was the long-term goal of those in the logistics community. Program Managers and influential senior logistics workers had long sold the dream of a unified system where all parts could be managed for the DoD. The status quo had each individual agency maintaining their own logistics system. These systems were a combination of government owned and sustained systems as well as commercial entities that were contracted out for government use. The software suites also had different funding streams. Some were program funded, and others funded at a working level. Due to this, there was not one governing body that could direct how these logistics systems operated. Without centralization, the agencies performed their logistics functions with varying levels of success. For these and other reasons, it was finally decided that a centralized system was needed.

Normally, when undertaking a large engineering effort, programs will bring senior engineering leaders and Chief Engineers onboard early to assist with the planning of the software. For an unknown reason, during the initial planning process for the ELSS system, no engineers, software or otherwise, were onboarded to the process. Therefore, the requirements for the system and the program planning were all performed by career logisticians. This is important to note that due to this, there were many mistakes in the early days of lifecycle planning and the first engineering efforts. As there were no engineers associated with the project, requirements are scarce. There were a few dozen very high-level requirements set down for contract bid. There was no mention of required MBSE or DoDAF efforts. The program was setup to follow the old Major Acquisition Pathway versus the Defense Business System Pathway of the modern Adaptive Acquisition Framework, as is required by law. Finally, contracting was to be through a Firm Fixed Price contract, which is much more difficult to perform Agile software development in than one under a Cost-Plus model.

Finally, after over a year of programmatic development, a systems engineering was brought into the project to offer guidance and assistance. As the contract was already put out to bid, there wasn't anything that could be done at first about the contract issues and acquisition pathway problem. There was, however, the ability to course correct on the architecture and requirements planning. This was where the systems engineering focused their efforts in the early stages. Documentation and policy were laid out for the use of MBSE and to the understanding that as a Major Acquisition Pathway project, DoDAF diagrams must be part of the delivery. The software needed to be developed in an Agile or Hybrid model. Requirements needed to be much more robust, detailing as much as could be done as possible from existing systems that were to be absorbed into the new ELSS. There was a lot of work to get done and not a lot of time to do it, as contractors were already submitting bids for the potential system.

6.6.3 Contract Competition

By 2022, the Program Manager decided that two contractors would compete for the ability to develop the ELSS system. This decision was based upon the "normal" operation of the Major Acquisition Pathway of the DoD Acquisition policy. The systems engineer was again not consulted on this decision, as this is not a normal execution for a software system. The two contractors were informed that they each had $25 million dollars to generate a more robust proposal, including a 5-year development schedule, high-level data architecture, development methodologies, and supporting documentation.

When the systems engineer heard of this, there was an obvious level of frustration. It is not common to have contractors compete in a software development project in this manner. There was no prototype to deliver, according to the contract. The amount of money being allocated to the effort of $50 million dollars could have resulted in a functional initial prototype, according to the systems engineer. The program was set up to be developed in the same methodology as would be used to procure a hardware system. This rigorous pathway works well for those types of systems. It does not, however, work well for software systems which do not need to perform logistics planning for the development and sustainment of the system, nor manufacturing planning, field exercises, etc.

The systems engineering noted that most of the requested documentation to be developed was useless for a software system. For example, each contractor had to deliver a Life Cyle Sustainment Plan (LCSP). As detailed in the DoD Directive 5000.01, an LCSP is "a comprehensive strategic document that details how a system, particularly a weapon system or IT solution, will be supported and maintained from its initial concept and development through its entire operational life and eventual disposal" [6]. These plans are created for "covered system", or systems that are national defense critical, and for weapons systems. A defense business system is only required to develop a Capability Support Plan (CSP) which "outlines the desired end-state, analyses of problems, and the future capabilities required to achieve the desired business mission, providing a roadmap for acquiring and evolving the system to meet strategic goals" [12]. A Life Cycle Sustainment Plan (LCSP) is a specific, required document in the Department of Defense (DoD) that details how a military system will be supported from its conception to its disposal. In contrast, a Capability Support Plan (CSP), as defined in the DoD's Adaptive Acquisition Framework for business systems, describes the governance and management approach for a business capability, focusing on resources and prioritization of sustainment activities after a capability is fielded. Therefore, an LCSP should never have been developed for the ELSS. This is just one example. There were numerous other problematic underutilizations of the contractors and wasted efforts as the program was developed under the same engineering standards as would be used for a hardware weapons system.

At the end of the contract competition, each contractor delivered the required documentation. A down select committee evaluated the contractors against the contract selection criteria and a winner was selected. The winning company then moved forward into a development phase.

6.6.4 ELSS Development Phase

The contractor and Program Office had no small task ahead of them. The project was now on its third program manager. The new Program Manager stood up a steering committee to act as their core advisors. The committee consisted of another Logistics Program Manager, a senior DoD Logistician, a Supply Chain Risk Management expert, and a Chief Engineer. The Chief Engineer role was filled by the previous systems engineer. This team worked diligently to bring the program into alignment with more Agile practices and the requirements of the Adaptative Acquisition Framework.

Luckily for the program, the contractor was fully onboard with a hybrid Agile approach for development. It was decided that the SEFAD method would work well for the development of the system. As the requirements were not fully detailed, the program entered the SEFAD pathway at Phase I, reworking the requirements and needs. Since time was important for the program and initial needs and requirements had already been developed, Phase II took place concurrently with the operation level of system model being developed at the same time. The diagrams developed for the system model acted as a driving force for discussions and understanding of the problem domain, allowing for more refined requirements to be developed and a better understanding of the schedule required for specified system functions. This became very clear during a discussion about the difficulty loading a bill of materials into the database. The contractor did not fully grasp the complexity of the problem as they had no experience in the performance of this task. The government did a poor job initially communicating the steps of the process. The contractor made a SysML Activity Diagram, explaining the steps they thought would have to occur. This allowed a dialogue to arise between the contractor and the government agents, detailing where the contractor underestimated the efforts. For example, the contractor had thought that bills of material would be provided by companies in a standardized format. This was an incorrect assumption as there is no standardized format for bills of material. Rather, in the normal process, a worker must transcribe whatever documents are provided to the government into a fully connected tree utilizing a spreadsheet. This allows for many errors, especially when considering recursive calls being introduced into the system during data load. Therefore, there must be an error checking phase and continuous monitoring of the load cycle. Through discussions fueled by MBSE diagrams, an acceptable solution was created for this problem, along with many others. This shows the true power of MBSE when combined with system development in that words are hard to visualize, diagrams are not, and humans are first and foremost, visual creatures. The creation of diagrams takes time to perform but results in a much more well engineered solution.

Contracting was also an issue that had to be addressed. With concurrence from the contractor, the Program Manager and Contracting Officer performed dozens of contract modifications, adjusting deliverables and contract requirements to allow for SEFAD implementation. The contract could not be fully changed, however, and the contractor did absorb some of the risk. This was done willingly and showed a great deal of trust between the contractor and the Program Office. Each side had to accept the risk that the other side would not deliver but trusted that they would. In this case, the ELSS development team has been very successful in working together.

As was noted earlier, the development pathway that was initially selected was the previous versions Major Acquisition Framework. The original Major Capability Acquisition Process required quite a few Systems Engineering Technical Reviews (SETR) events, included many go/no-go gateway decisions, milestone decisions, required and inapplicable documentation generation, and little to no early looks at the product, as shown in Fig. 6.9. This sort of development path is very common in the Predictive world and did not reflect the Agile methods being used in the development of the ELSS. As this process was also outdated and had been replaced with the new Adaptive Acquisition Framework, the Program Manager and Chief Engineer spent a good amount of time adapting the project execution to fit the Defense Business System and Software Development Pathways. These pathways do away with the cumbersome and unnecessary SETR events, focusing instead on capability development in a collaborative and iterative manner, as shown in Fig. 6.10. As the program was still stood up under the "Major Capability Acquisition" pathway, certain milestones still needed to be held, but the program was able to nest the Defense Business System and Software Acquisition pathways within the Major Capability Pathway, thereby increasing the Agility of the project execution.

A topic of interest in the development of this system is that at the start of the development, the COVID-19 pandemic had occurred, and the government team was fully teleworking. The various agency representatives, steering committee members, and contracting personnel were spread out all around the USA, making the development of the system effectively a fully remote effort. With a 3-h difference between East coast and West coast in the USA, working hours were flexible. Meetings were scheduled to be early morning and midday for the West coast participants, allowing for meetings to be in the afternoons for the East coast. Utilizing Agile methods allowed the development team to work their own hours, checking in

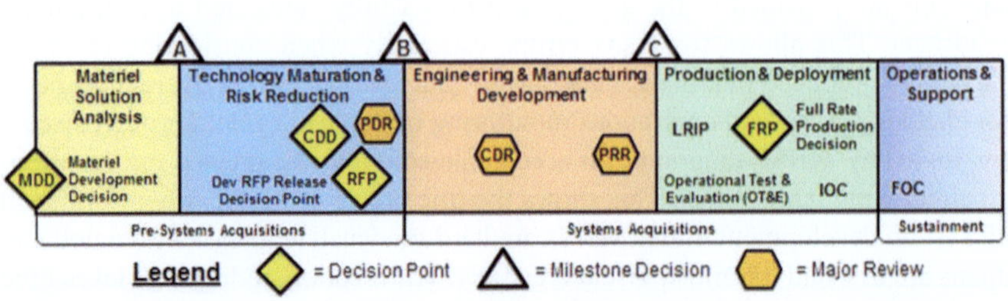

Fig. 6.9 Major capability acquisition process (2015) [6]

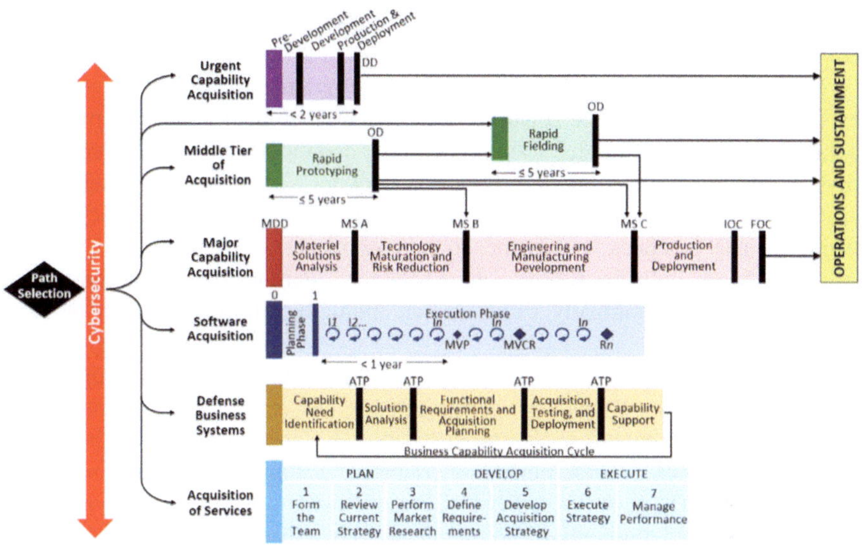

Fig. 6.10 Adaptive acquisition framework pathways (2020) [3]

daily at the Scrum and reporting to the steering committee at the Sprint Reviews. Development execution and morale was high with the developers citing the methodology as one of the reasons behind it.

6.6.5 Conclusions and ELSS Future

The ELSS case study demonstrates the transformative impact of integrating systems engineering discipline with Agile adaptability through the SEFAD framework. What began as a fragmented and poorly structured enterprise initiative evolved into a coordinated, requirements-driven, and model-informed development effort that successfully aligned technical execution with mission intent. The introduction of formal systems engineering practices, particularly the adoption of MBSE and the incorporation of SEFAD's structured Agile principles, provided the clarity and rigor necessary to rescue a program that had initially suffered from inadequate planning and misaligned acquisition strategy.

One of the defining lessons from ELSS is the importance of early systems engineering involvement. The absence of engineering leadership in the program's formative stages resulted in unrealistic requirements, inappropriate acquisition pathway selection, and contractual inefficiencies. Once a systems engineer and later a Chief Engineer were embedded into the program, course correction became possible. The shift toward MBSE modeling allowed for deeper understanding of operational workflows, revealed hidden system complexities, and improved communication between technical and functional stakeholders. This visual and iterative approach not only enhanced requirements definition but also built mutual trust between the contractor and government representatives.

The successful collaboration between the Program Office and the contractor exemplifies how risk can be constructively shared when both parties operate under a transparent, SEFAD-guided framework. Agile ceremonies provided a predictable rhythm of communication and progress tracking, while systems engineering artifacts-maintained traceability and verification confidence. By adopting hybrid Agile-SEFAD practices within the boundaries of the Department of Defense's Adaptive Acquisition Framework, the ELSS team demonstrated that compliance and agility are not mutually exclusive but mutually reinforcing when properly structured.

Equally noteworthy is the program's resilience in the face of remote operations during the COVID-19 pandemic. Distributed teams across multiple time zones maintained high engagement and productivity, a testament to the effectiveness of SEFAD's emphasis on communication cadence, accountability, and empowered team structures. The use of daily Scrums, Sprint Reviews, and iterative development milestones ensured momentum and morale, even in a dispersed environment.

In summary, the ELSS program validated SEFAD's premise that speed and rigor can coexist within complex defense software systems. The introduction of systems engineering leadership, MBSE visualization, and Agile collaboration reshaped a high-risk enterprise into a disciplined, adaptive, and high-performing program. The success of ELSS should serve as a model for future Department of Defense software initiatives, demonstrating that integrating engineering rigor early and sustaining it through continuous Agile adaptation is essential to achieving mission assurance in modern digital acquisition. As of the time of this textbook publish, the ELSS system is still in development, but it shows high potential for successful deployment.

6.7 Case Study 7: How a Lack of Engineering Rigor Led to the Crowdstrike Crisis

6.7.1 Introduction

In July 2024, Crowdstrike, a cybersecurity company that provides an AI-powered, cloud-native platform to protect endpoints, cloud workloads, identity, and data from cyberattacks, pushed a release version of software that caused a catastrophic failure throughout the company's user base. A routine update to Crowdstrike's Falcon endpoint protection agent triggered a Blue Screen of Death (BSOD)[6] critical error on millions of Windows based machines worldwide, with some reports stating over

[6] A Blue Screen of Death (BSoD) is a fatal error screen in Microsoft Windows that appears after a critical system error, indicating a problem that the operating system cannot recover from and forcing a system shutdown or restart to prevent further damage.

24,000 customers impacted, causing unprecedented disruptions within multiple critical systems used by airlines, finance, defense, the medical industry, and more. In the airline industry alone, it is estimated that over 1000 flights were canceled due to the botched Falcon rollout. This catastrophe illustrates how insufficient engineering and technical rigor in critical systems can create a network cascade of systemic failures. In this section, we will analyze how this failure happened, exploring the problems realized by Crowdstrike through three lenses: procedural and process failures, technical and architectural lapses, and organizational and cultural pressures.

6.7.2 Procedural and Process Failures

One of the core deficiencies noted was Crowdstrike's verification and validation activities. Crowdstrike utilized a CI/CD pipeline and relied heavily upon automated unit testing. The defective update that slipped through the automated testing was a configuration change and was deployed broadly to their customers. This ultimately caused out-of-bounds memory access and system crashes. Because this configuration change interacted at the kernel level due to the Falcon software suite being tightly integrated with the Windows operating system, the defective update became a catastrophic problem as any machine that loaded the software crashed the Windows operating system.

In systems with high-quality and low-defect requirements, as all critical systems are by definition, test coverage and validation must include boundary conditions, negative tests, and automated safety checks, such as fuzz testing and memory safety analysis. In the case of Crowdstrike's defective Falcon publish, the Falcon rule interpreter configuration parser apparently accepted malformed data or wildcard patterns without adequate constraint checking. Furthermore, the update seems to have been deployed automatically to their large user base, without a staging or "canary" rollout[7] implemented, causing a multiplying risk factor.

Due to the faulty module being deeply ingrained in the system boot, machines became stuck in an infinite boot loop and could not download and apply the attempt automated patch fix. This resulted in the requirement of manual intervention, such as booting into Windows safe mode, manually deleting the offending file, etc. With potentially hundreds of thousands of machines affected worldwide, this was an immense task.

If Crowdstrike had utilized a development methodology and process focused on engineering and technical rigor with stronger rollback and fail-safe mechanisms rather than an overly Agile and flexible stance, this crisis might have prevented the window of disruption from propagating so broadly. For example, if versioned configurations, safe fallback defaults, or dual-image deployment had been used, the

[7] A canary rollout is a software deployment strategy that releases a new version or feature to a small subset of users or servers before a full rollout to the entire user base.

damage might have been contained. Unfortunately for Crowdstrike and the respective customers, such rigorous fallback paths were apparently either absent or ineffective.

6.7.3 Technical and Architectural Lapses

The Crowdstrike Falcon application runs at a privileged level of the Microsoft Windows Operating System. By running at the kernel level, Falcon has strong monitoring power. This level of operation also comes with high systemic risk as failure in Falcon can result in a failure of the operating system itself, as was shown in the 2024 catastrophe. In systems engineering, we call this a tightly coupled architecture. A more robust design would have partitioned the high-trust components and decouple the configuration handling away from the critical path of the boot operations. In this case, the update path was on the critical boot operation path, and the system lacked isolation strong enough to absorb errors.

It appears that the configuration file parsing allowed wildcard, or regex patterns, loaded them into arrays, and failed to properly bound-check the indices. In a more engineering rigor focused regime, configuration formats and parsers must have formal textual grammar, incorporate strict schema validation procedures, and be contain bounds checking. Changing a configuration should be treated the same as changing code, with all associated unit tests, parser regressions, and safety invariants, not to mention formalized manual functional testing, both for regression and edge cases. Also of note, only valid data was tested in the manual testing that was performed.

Of note, internal tools for content validation utilized in Crowdstrike's content validation software reportedly had bugs of its own. The validation tool allowed invalid patterns to pass through. The content validation software did not recognize that a mismatched number of inputs had been received and that it would cause a systems crash. This resulted in what looked like a safe configuration to pass existing checks but trigger runtime failures.

Because so many enterprise systems were reliant on Microsoft Windows and Crowdstrike's Falcon software, the failure was massively amplified and resulted in a reduction of resiliency. A weaker design would anticipate that any failure in a widely deployed component must be extremely robust, tested, small stage-deployed, and have a throttled rollout. The systemic homogeneity became a multiplier of risk. In systems engineering, this is a harsh reminder that components which lie on the shared infrastructure backbone of a system must adhere to the highest levels of engineering and technical rigor as faults in these components quickly become system wide problems.

6.7.4 Organizational and Cultural Pressures

Before the breach, CrowdStrike had reportedly implemented workforce reductions that affected engineers, developers, and software testers. At first glance, such measures may appear to improve short-term efficiency or reduce costs, but in the context of critical software systems, they carry hidden risks. When experienced personnel are removed, institutional memory, or the accumulated knowledge of past design decisions, lessons learned from prior incidents, and informal but vital "engineering intuition," is diminished. This weakens the organization's ability to anticipate failure modes that may not be fully documented but are nonetheless known to seasoned staff.

Reducing staff decreases redundancy in the review process, weakening one of the fundamental safeguards of engineering rigor. Critical systems that demand high-quality assurance often rely heavily on multiple independent evaluations of developer committed design, code, and configuration changes to catch often unseen but dangerous flaws. When staffing levels are cut, there are fewer reviewers available to provide this essential oversight. This results in a far increased probability of defects and errors in software code commits. Such vulnerabilities are especially dangerous in critical systems, where even a single overlooked defect can propagate rapidly across interconnected networks and infrastructures, resulting in cascading failures throughout multiple systems. In critical systems, the margin for error is already minimal, and reducing review capacity compromises the very resilience such systems are supposed to guarantee.

The most dangerous aspect of these dynamics is that the weakening of engineering rigor is often obfuscated until crisis occurs. On ordinary days, the system may appear stable and performant, giving leadership a false sense of security. However, without robust staffing, the underlying safety nets maintained through reviews, regression tests, configuration checks, and post-mortem analyses are compromised. It is only in the presence of a catastrophic event, such as the CrowdStrike outage, that the consequences of reduced engineering capacity become visible.

From a systems engineering standpoint, this is a clear example of the danger in trading resilience for efficiency. Cutting staff may look good on a balance sheet in the short term, but it directly undermines the system's ability to absorb shocks and recover when things go wrong. In critical domains, those cuts aren't just financial decisions. Rather, they reshape the entire risk posture of the organization. This loss of rigor doesn't happen all at once. It builds slowly as checks are skipped, feedback loops weaken, and institutional knowledge disappears. That erosion creates conditions where small defects slip through unnoticed and build into larger vulnerabilities. When those vulnerabilities surface, the lack of resilience mechanisms turns what could have been a contained issue into a cascading, system-wide failure, which is unacceptable for critical systems.

6.7.5 *Crowdstrike Falcon Conclusions*

The Crowdstrike Falcon crisis demonstrates how lapses in the implementation and maintenance of engineering and technical rigor can spiral into global disruption for critical system. Even the smallest routine actions in these types of system can have large impacts overall. In this case, the chain of failure included weak verification and validation activities, deficient rollback policies and procedures, unnecessarily tight architectural coupling, and internal cultural pressures that eroded existing safety margins. Critical software systems demand engineering rigor as a core necessity. The Crowdstrike failure reveals that even in mature and highly trusted market-leading companies, the smallest lapse in discipline can propagate to a catastrophic scale.

6.8 Case Studies: Conclusions

What are the systemic causes for success and failure of DoD software projects? After reviewing multiple DoD software project executions, it is readily apparent the problem that defense development teams face. Software is often developed without good requirements management and is grown from smaller, proof of concept type projects, or are systems that have been supported for decades and have little documentation. Due to the nature of software development, teams can fluctuate in size as projects go through growth and diminishment, especially during resource droughts. DoD software projects can last years, with developers leaving and returning projects. As opposed to Predictive Projects, many of these projects are executed in an Agile Scrum style project management and there is little documentation available. Systems that are decades old tend toward Predictive methods with lip service given to Agile development. Those developing under Agile Scrum run the gamut for little to no actual Agile Scrum alignment and extreme engineering rigidity bordering on Predictive execution to those who have no requirements, documentation, or any level of engineering or project rigor where developers have no oversight and work on what they want, when they want, with no planning. Due to this, knowledge management and knowledge transfer is low, resulting in teams operating in a “tribal knowledge” style situation, with senior developers passing on knowledge verbally to junior developers. These problems can potentially be fixed by thoughtful implementation of systems engineering rigor with the Agile Scrum development standard.

6.9 Discussion Questions

Implementing structured requirements in the defense data exchange software system (DESS): How did the introduction of structured requirements management in the DESS project alter the team’s engineering discipline and communication culture, and what lessons can be drawn about overcoming resistance to formal rigor within Agile development teams?

Implementing systems engineering rigor in the defense logistics management system (DLMS): In what ways did the DLMS team's adoption of systems engineering practices such as baseline management, traceability, and verification improve the quality and predictability of deliverables, and how can similar practices be scaled across other Agile defense programs?

Project history of the DoD process automation system (DPAS) 3.0 conversion: What does the DPAS 3.0 case reveal about the consequences of excessive agility without systems engineering control, and how did leadership reintroduce structure to recover program stability and ensure mission assurance?

How too much rigor can stifle innovation in the defense qualification management system (DQMS): How can project leaders identify when systems engineering rigor has crossed the threshold from productive control to innovation suppression, and what strategies might allow a program to retain compliance while restoring creativity and responsiveness?

Adapting a work management system to modern software and systems engineering practices (DWMS): How does the DWMS case illustrate the risks of partial implementation of Agile Scrum and systems engineering rigor, and what specific corrective actions such as role realignment, story point reform, and formal requirements documentation would most effectively bring the program into compliance with modern SEFAD and INCOSE standards?

Successes in designing the enterprise logistics software system (ELSS) utilizing the SEFAD Process: How might the outcome of the ELSS program have differed if systems engineering principles—such as requirements traceability, model-based systems engineering (MBSE), and acquisition pathway alignment—had been applied from the program's inception? In your answer, consider how early systems engineering engagement could have influenced contracting strategy, stakeholder communication, and overall development efficiency.

How a lack of engineering rigor led to the CrowdStrike crisis: How did the CrowdStrike crisis exemplify the systemic dangers of neglecting engineering rigor in verification, architectural design, and organizational culture, and what specific safeguards—such as staged rollouts, stronger configuration validation, and preservation of institutional engineering expertise—could have prevented or contained the global cascade of failures?

Comparative analysis of case studies: When analyzing all seven case studies together, what common patterns emerge regarding the interplay between flexibility and discipline, and how might SEFAD serve as a framework for balancing these forces across large, multi-domain defense projects?

Lessons learned from case studies: Which specific leadership behaviors, communication strategies, or cultural adaptations most strongly influenced the success or failure of SEFAD adoption across the case studies, and how can these insights inform future organizational transformation initiatives?

The path forward: Institutionalizing SEFAD Across Programs: What institutional, technical, and cultural steps must an organization take to fully embed SEFAD principles into enterprise-level policy and practice, ensuring that agility and rigor remain harmonized across evolving missions and technologies?

References

1. Defense Acquisition University, Test Readiness Review (TRR) (Defense Acquisition University, 2023). [Online], https://aaf.dau.edu/aaf/mca/trr/#:~:text=A%20Test%20Readiness%20Review%20(TRR,and%20gathers%20the%20required%20information
2. Defense Acquisition University, Production Readiness Review (PRR) (Defense Acquisition University, 2023). https://aaf.dau.edu/aaf/mca/prr/. Accessed 12 May 2023
3. Office of the Under Secretary of Defense for Acquisition and Sustainment, DOD Instruction 5000.02: Operation of the Adaptive Acquisition Framework, Washington (2022)
4. Kumar, S. A Review on Client-Server Based Applications and Research Opportunity, International Journal of Recent Scientific Research, vol. 10, no. 7(H), pp. 33857–33862, 2019.
5. Jørgensen, M. (2018). Do Agile Methods Work for Large Software Projects?. In: Garbajosa, J., Wang, X., Aguiar, A. (eds) Agile Processes in Software Engineering and Extreme Programming. XP 2018. Lecture Notes in Business Information Processing, vol 314. Springer, Cham. https://doi.org/10.1007/978-3-319-91602-6_12
6. Office of the Under Secretary of Defense for Acquisition and Sustainment, DoD Directive 5000.01: Operation of the Defense Acquisition System (2015). https://www.acq.osd.mil/fo/docs/500002p.pdf. Accessed 17 Aug 2024
7. Assistant Secretary of Defense for Mission Capabilities, PDR and CDR Assessments (Office of the Under Secretary of Defense for Research and Engineering, 2024). https://ac.cto.mil/dte/pdr-cdr/. Accessed 17 May 2024
8. Office of the Under Secretary of Defense for Research and Engineering, *Systems Engineering Guidebook* (Office of the Under Secretary of Defense for Research and Engineering, Washington, 2022)
9. Office of the Under Secretary of Defense for Acquisition and Sustainment, Software Acquisition, Washington, DC: Office of the Under Secretary of Defense for Acquisition and Sustainment, 2022.
10. Office of the Under Secretary of Defense for Acquisition and Sustainment, *DOD 5000.87: Operation of the Software Acquisition Pathway* (Office of the Under Secretary of Defense for Acquisition and Sustainment, Washington, 2020)
11. DoD Chief Information Officer, *DoD Enterprise DevSecOps Strategy Guide* (Department of Defense, Washington, 2021)
12. Office of the Under Secretary of Defense for Acquisition and Sustainment, *DoD Instruction 5000.75: Business Systems Requirements and Acquisition* (Office of the Under Secretary of Defense for Acquisition and Sustainment, Washington, 2020)

Chapter 7
Rigorous Software Quality Assurance

7.1 Verification and Validation in a Systems Engineering Environment

7.1.1 Verification and Validation Introduction

Verification and Validation, often abbreviated as V&V, represent two of the most essential activities associated with well-engineered systems within the practice of systems engineering. In the context of software-intensive critical systems, these activities form the foundation of confidence that a system will perform safely, securely, and reliably in the environment for which it is intended. According to INCOSE, verification is the process of determining that the system has been built correctly in accordance with its specified requirements, while validation ensures that the right system has been built to satisfy the intended mission and stakeholder needs [1]. When either discipline is neglected, the result can be catastrophic. In critical systems such as command-and-control networks, autonomous defense platforms, nuclear monitoring systems, and medical device software, even a single undetected defect can lead to operational paralysis or the loss of human life. For this reason, verification and validation must not be treated as final project activities but as continuous and integral elements of the entire system lifecycle.

7.1.2 Verification: Building the System Right

Verification is the formal process of confirming that every requirement, design element, and implemented function of a system meets its intended specifications (Fig. 7.1). In software-intensive environments, verification must be conducted through structured and repeatable processes such as analysis, test, inspection, and

D. Rosson, *Systems Engineering in Agile Software Development*,
https://doi.org/10.1007/978-3-032-14613-7_7

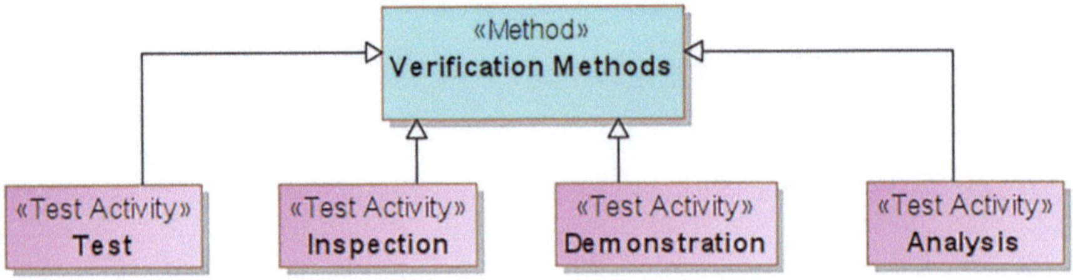

Fig. 7.1 Verification methods

demonstration. INCOSE identifies requirement traceability as a cornerstone of effective verification, ensuring that every "shall" statement is directly linked to its design representation, implementation artifact, and the evidence that confirms compliance [1]. Through this traceability, full system and problem domain understanding is achieved. Simply put, the quality assurance personnel have a much easier time defining, developing, and delivering successful verification of system function when requirements are well written and traced throughout the system architecture and lifecycle. Verification is not limited to functional correctness but must also account for non-functional requirements including safety, security, performance, reliability, and maintainability.

For example, in a real-time embedded aerospace application, verifying functionality may involve confirming that sensor inputs trigger the appropriate system response. However, verifying non-functional behavior requires proving that the system meets latency constraints, that memory allocation does not cause timing jitter, and that data transfer remains secure under all operational conditions. These verification activities must be documented through objective evidence that can be independently reviewed and repeated. Modern systems engineering environments achieve this using automated test frameworks, continuous integration pipelines, and simulation tools that provide continuous feedback throughout development.

Verification within software-intensive critical systems also extends into the domain of configuration management and version control. Every release of the software must be baselined, tested, and validated against regression test suites to ensure that no previously verified function has been compromised. Formal peer reviews, static code analysis, and automated safety checks such as fuzz testing and boundary analysis are vital tools that strengthen verification rigor. The goal of verification is to establish measurable confidence that every requirement has been fully implemented and that the system operates within its defined performance boundaries under all foreseeable conditions.

7.1.2.1 Goals of Verification Activities

The goal of software verification is to prove, with objective evidence, that the software correctly and completely implements the specified requirements while ensuring that the product is technically sound, compliant, and ready for validation by stakeholders.

- Confirm Compliance with Requirements:
  - Ensure that every requirement (functional and non-functional) has been implemented as specified.
  - Demonstrate that all "shall" statements have corresponding test cases or verification evidence.
  - Verify traceability between requirements, design, code, and test results.
- Ensure Technical Correctness:
  - Evaluate whether the software performs as designed under defined conditions.
  - Identify defects in algorithms, logic, and interfaces before deployment.
  - Confirm that no unintended behavior is introduced through design or coding errors.
- Assess Implementation Completeness:
  - Verify that all planned functionality, data handling, and interfaces are fully realized.
  - Ensure that all software components integrate properly and communicate as intended.
  - Confirm that configuration items are complete and baseline documentation matches the delivered system.
- Demonstrate Quality and Performance:
  - Confirm that the system meets performance thresholds (speed, memory usage, response time, etc.).
  - Assess robustness, stability, and error-handling mechanisms.
  - Evaluate maintainability, portability, and scalability where required.
- Provide Objective Evidence for Certification and Audit.
  - Produce verifiable, repeatable test results that demonstrate conformance to contractual, regulatory, and safety standards (e.g., DoD 5000.88, IEEE 1012, ISO/IEC/IEEE 15288).
  - Establish formal documentation of test results, inspections, analyses, and demonstrations.
- Detect and Correct Defects Early:
  - Identify anomalies, deviations, or mismatches before operational testing or validation.
  - Reduce downstream rework and risk through iterative and continuous verification within each sprint or increment.
- Maintain Configuration and Traceability Integrity:

- Ensure that changes to requirements or code are properly captured, versioned, and verified through updated test cases.
- Preserve bidirectional traceability through the digital thread: Requirement → Design → Code → Test → Verification Evidence.

- Support Continuous Integration and DevSecOps Goals:
  - Embed verification into Agile cycles (e.g., automated unit and integration tests in CI/CD pipelines).
  - Provide real-time feedback on system quality, ensuring that every increment maintains compliance and readiness.

7.1.3 Validation: Building the Right System

While verification ensures correctness against specification, validation ensures relevance to stakeholder intent. According to INCOSE, validation demonstrates that "the realized system satisfies the intended use when placed in the operational environment" [1]. In software-intensive, critical systems, this requires close interaction between developers, users, and domain experts to confirm that the implemented capabilities meet mission objectives under realistic conditions. Validation scenarios often employ simulation environments, digital twins, or hardware-in-the-loop testing to replicate operational contexts that are too hazardous, costly, or complex to test directly.

A method commonly used for validation is customer acceptance testing (Fig. 7.2). Customer acceptance testing is the act of granting the end user the ability to access the system prior to a software release, having full access to use the system under normal operational conditions and providing feedback for the developers. This is most successful when the software user is available for testing and there is a good level of communication between the software development team and the customer. The other key driving factor of success in customer acceptance testing is giving access to early builds based upon Sprint Review iterations, allowing customers to have "hands on" experience with the product prior to final acceptance. This allows the development team to make use of the "feedback" and "continuous learning" ways, pivoting and adjusting as customers report successes and failures.

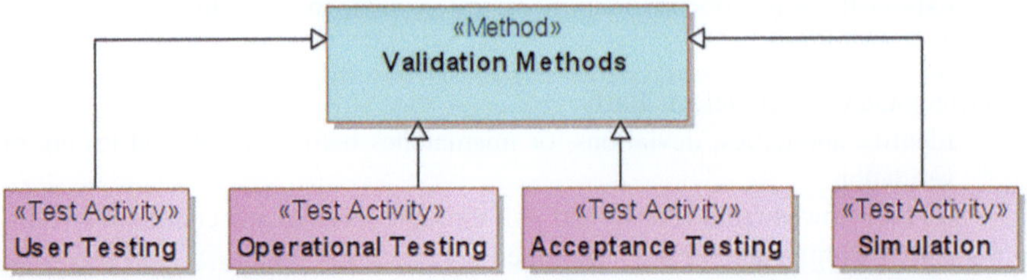

Fig. 7.2 Validation methods

The challenge in validation lies in bridging the gap between technical performance and mission effectiveness. A validated logistics optimization algorithm, for example, must not only generate correct outputs but also integrate seamlessly with human workflows, data sources, and security constraints. INCOSE guidance emphasizes stakeholder engagement, iterative assessment, and mission modeling as essential elements of an effective validation strategy [1]. For critical systems, validation must also include operational safety and assurance reviews such as structured analyses that confirm the system's behavior will not cause unintended harm when deployed in real environments.

7.1.3.1 Goals of Validation Activities

Software validation ensures that the system fulfills its intended use in its real-world context. It confirms that user and mission needs are satisfied, the system performs as expected in its operational environment, and that the delivered capability provides the intended value.

- Confirm stakeholder and mission needs are met:
  - Ensure that the software solves the real operational problem it was intended to address.
  - Demonstrate that the system capabilities align with stakeholder intent, user expectations, and business or mission outcomes.
  - Validate that system behavior supports end-to-end workflows and decision-making as envisioned by end users.
- Evaluate the system in its operational environment:
  - Test the software under realistic or simulated field conditions (including load, latency, and environmental factors).
  - Confirm interoperability with external systems, hardware, or data sources in the target environment.
  - Identify context-specific usability or performance gaps that might not appear in laboratory testing.
- Ensure the solution is "fit for purpose":
  - Assess whether the system, as delivered, is useful, usable, and effective in its intended domain.
  - Determine that it supports mission readiness, user workflows, and operational tempo.
  - Validate that design trade-offs (cost, performance, security, reliability) are acceptable to stakeholders.
- Build stakeholder confidence:
  - Demonstrate to users, customers, and oversight authorities that the software meets its mission objectives.
  - Present measurable evidence of capability through acceptance testing, user evaluations, or operational demonstrations.

- Establish the trust necessary for deployment, certification, or fielding.

- Detect and resolve misalignment between requirements and user intent:
  - Identify when technically correct (verified) features fail to meet user expectations or operational needs.
  - Capture feedback to refine requirements, workflows, and interfaces.
  - Prevent the delivery of software that is compliant on paper but ineffective in practice.

- Support continuous feedback and improvement:
  - Integrate validation into each Agile Sprint or Increment, not just at the end of the lifecycle.
  - Gather user feedback through Sprint Reviews, demos, and field tests to refine capability iteratively.
  - Align validation findings with requirement updates to maintain system relevance and adaptability.

- Provide evidence for acceptance and deployment:
  - Document that the system achieves Initial Operational Capability (IOC) or equivalent readiness milestones.
  - Supply validation artifacts required for certification, accreditation, or authority to operate (ATO).
  - Enable informed decision-making for release, sustainment, or improvement phases.

7.1.4 Integrated V&V Across the Lifecycle

In traditional engineering environments, verification and validation are often treated as end-of-cycle activities that occur after the system is built. This approach is no longer viable for modern, software-driven critical systems. The contemporary systems engineering paradigm, as endorsed by INCOSE and reflected in Systems Engineering Focused Agile Development (SEFAD), integrates V&V throughout the entire development lifecycle. Each sprint, iteration, and design review includes defined verification checkpoints and validation activities. This continuous approach ensures that both correctness and mission alignment are maintained as the system evolves.

Integration of verification and validation throughout the lifecycle requires both cultural and technical adaptation. Technically, it demands traceability between requirements, architecture, design artifacts, and test results. Culturally, it requires engineers, developers, and program leaders to view quality as a constant responsibility rather than a final milestone. Testing is not pushed off to the end of the Sprint but rather handed downstream as development is marked as "complete." By embedding V&V into the daily workflow, organizations reduce the cost of defect discovery and significantly increase confidence in system reliability.

Continuous integration and continuous deployment pipelines, often referred to as CI/CD systems, are central to this approach. Each commit to the code repository triggers automated unit, integration, and system-level verification activities, while model-based simulations and digital engineering environments provide real-time validation feedback. Early prototypes and simulated operational environments serve as testbeds where both verification data and validation insights are captured concurrently. This not only shortens feedback loops but also improves stakeholder understanding of progress and risk posture.

Ultimately, embedding verification and validation into the daily rhythm of engineering work fosters an environment of shared accountability and technical excellence. It ensures that agility never compromises assurance and that the pursuit of rapid capability delivery is always grounded in disciplined engineering practice. In the domain of critical systems, where failure is unacceptable, continuous V&V stands as both a safeguard and a catalyst—protecting mission integrity while enabling innovation to proceed with confidence.

7.1.5 The Imperative of Rigor in V&V

Verification and validation represent the highest expressions of discipline within the engineering process. They are not clerical or administrative activities intended merely to satisfy procedural compliance but rather, they are the operational mechanisms through which safety, reliability, performance, and ultimately trust is established in complex systems. The purpose of verification and validation is to ensure that a system is both built right and built for the right purpose. This dual responsibility demands not only technical precision but also organizational maturity. In practice, this means that V&V must be treated as fundamental engineering functions, fully integrated into planning, design, and execution, and given the same attention, resources, and authority as system architecture or implementation.

In critical software systems, the consequences of neglecting verification and validation can be severe and far-reaching. When a flight control algorithm, a power distribution scheduler, or a cybersecurity agent is inadequately verified, the resulting defects can propagate invisibly through interconnected architectures, compromising entire mission chains. This is particularly true in modern networked environments where systems no longer operate in isolation, and the Modular Open Systems Approach (MOSA) and Systems of Systems are far more commonplace. A small configuration error or unverified boundary condition in one component can cascade through dependent systems, resulting in widespread mission degradation or total operational failure. Rigorous verification and validation act as the systemic immune response, detecting and neutralizing these defects before they are allowed to manifest in the field.

INCOSE guidance emphasizes that V&V activities must be planned, resourced, executed, and independently reviewed with the same seriousness as system design and development [1]. This requires deliberate investment in test infrastructure, validation environments, and qualified personnel capable of executing independent assessments. Objectivity in verification and validation is critical for credibility. In safety-critical or mission-critical environments, internal confirmation by the development team is insufficient. Independent validation authorities, whether internal to the organization or through external certification bodies, must review and attest that the system performs correctly, safely, and securely under all defined operational conditions. This independent oversight provides both technical assurance and institutional accountability, reinforcing the integrity of the engineering process. In practice, this means that the adage of "if you touched the code, you don't test the code" is still highly applicable. Testing must be performed by another developer who has no experience with the code under test. Even better, the code should be tested by a professional software tester, removing any developer bias from the process.

History provides many sobering reminders of what happens when rigor in V&V is compromised. Failures in flight control systems, medical device software, and global cybersecurity rollouts have all demonstrated that even a single unverified or inadequately tested change can result in cascading system-wide consequences. In several high-profile incidents, insufficient boundary testing, incomplete validation of configuration data, and an overreliance on automation without human review led to global outages and loss of mission-critical functionality. Each of these failures underscores the same fundamental truth: in complex, interconnected systems, the cost of insufficient verification is not localized but exponential. The absence of rigor in one subsystem can destabilize the entire operational ecosystem.

The role of verification and validation in the engineering of software-intensive critical systems extends far beyond compliance. V&V serves as the guardians of mission assurance, providing objective evidence that the system not only performs as designed but also fulfills its intended operational purpose within its real-world environment. They ensure that innovation proceeds responsibly and that agility remains anchored in accountability. In a development culture increasingly focused on speed and adaptability, V&V anchor progress in measurable assurance, ensuring that every design decision is traceable to verified evidence rather than subjective confidence.

A well-executed verification and validation process builds organizational credibility. It assures stakeholders that system behavior can be predicted, audited, and trusted under both nominal and extreme conditions. For mission owners, this assurance is the difference between confidence and uncertainty. For end users, it is the difference between success and failure. It is the activity through which risk is mitigated. Ultimately, rigorous verification and validation transform complex engineering programs from fragile constructs of assumptions into resilient systems capable of withstanding uncertainty, evolving requirements, and the pressures of real-world operation. Through disciplined application of V&V, organizations demonstrate that engineering excellence is not measured solely by innovation but by the sustained ability to deliver safety, reliability, and performance in every operational context.

7.2 Applying Systems Engineering Principles in DEVSECOPS and CI/CD

7.2.1 Introduction

Modern critical software systems must evolve at the same pace as emerging threats, advancing technologies, and shifting policy mandates (Fig. 7.3). This environment of constant change demands an engineering approach that delivers capability continuously without sacrificing assurance. To meet this need, organizations have increasingly adopted DevSecOps, an integrated philosophy that merges development, security, and operations into a unified, collaborative process. DevSecOps extends beyond a set of tools or automation pipelines. It represents a cultural and procedural transformation that embeds security and verification throughout the software lifecycle. Central to this approach are the three ways of DevSecOps: the *flow* of work from development to deployment, the rapid *feedback* loops that allow detection and correction of defects, and the commitment to *continuous learning* and experimentation that fosters technical excellence and adaptability. Through these principles, DevSecOps supports continuous integration and continuous deployment (CI/CD), ensuring that every change is built, tested, and delivered automatically while maintaining the traceability, reliability, and resilience essential for critical mission systems.

While DevSecOps accelerates innovation, it also introduces unique risks when applied to critical systems, where any software defect or misconfiguration can have catastrophic operational effects. In this environment, systems engineering provides the necessary rigor, discipline, and governance to ensure that agility does not compromise assurance. As stated in the INCOSE Systems Engineering Handbook (Fifth Edition), systems engineering must "integrate technical and programmatic elements in a manner that assures mission success across the entire system lifecycle." Applying SE principles within DevSecOps ensures that even as delivery frequency increases, safety, reliability, and traceability remain uncompromised.

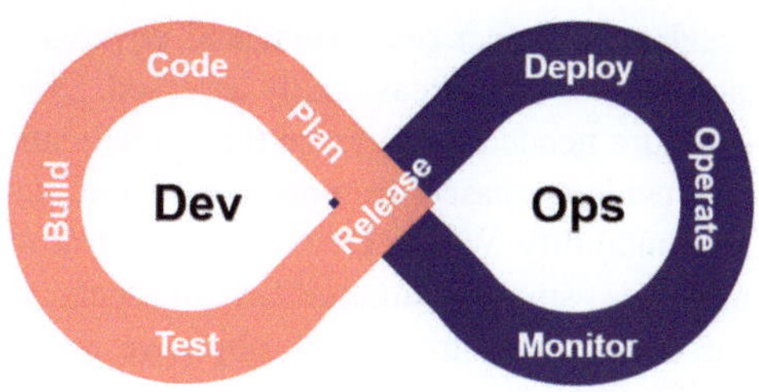

Fig. 7.3 DevSecOps lifecycle

7.2.2 Integrating Systems Thinking into Continuous Delivery

At its core, systems engineering introduces a holistic perspective that views the system as a dynamic, interconnected ecosystem rather than a sequence of isolated software components, versions, and releases. This mindset aligns naturally with the goals of CI/CD, which rely on the automated orchestration of builds, tests, and deployments across distributed environments. However, SE extends beyond automation and embeds structured traceability, verification, and risk management within every iteration of the pipeline.

In a properly engineered CI/CD environment, each code commit or configuration change is treated as a controlled change event. Systems engineers ensure that every modification can be traced to a requirement, that verification criteria are automatically executed, and that validation results are recorded as digital evidence. This aligns with INCOSE's emphasis on requirements traceability and verification evidence management as central to maintaining engineering rigor in complex systems. The use of digital engineering toolchains, including MBSE architectures, requirement repositories such as DOORS Next, and automated testing frameworks, allows verification artifacts to be linked directly to configuration baselines and deployment states.

7.2.3 Embedding Verification and Validation in the Pipeline

DevSecOps and CI/CD fundamentally transform where and how verification and validation occur within the engineering lifecycle. Traditional system development approaches often postponed testing until after major integration milestones, which created long feedback cycles and increased the risk that critical defects would remain undetected until late in development. In contrast, continuous integration requires that testing, verification, and validation occur automatically, continuously, and at multiple levels, ranging from unit testing to full system evaluation, throughout every phase of the lifecycle. This approach allows immediate detection of errors, accelerates defect resolution, and ensures that system quality is constantly measured rather than occasionally assessed. Systems engineering principles provide the structure needed to organize these verification activities into traceable, repeatable, and evidence-based frameworks that align with INCOSE's focus on rigor and accountability. Within DevSecOps environments, these frameworks integrate automated test suites, validation environments, and model-based simulations that generate continuous streams of objective performance data. The result is a living verification ecosystem in which every code change, model update, or configuration adjustment is validated against defined requirements, ensuring that reliability and mission assurance advance in harmony with development progress.

For critical systems, verification activities must include both functional and non-functional testing to ensure complete confidence in system performance and safety.

Functional testing confirms that each software component behaves as specified, while non-functional testing evaluates qualities such as reliability, security, scalability, and maintainability. Security scans, boundary condition analysis, and timing verification are essential for uncovering subtle errors that may not appear during routine operations but could cause catastrophic failures under stress. Automated test suites play a central role in this process by ensuring that every new software build is evaluated against predefined acceptance criteria before it is approved for deployment. These automated frameworks allow testing to occur continuously, providing immediate feedback to developers and engineers as code changes are introduced. Test Driven Development (TDD) drives the creation of unit tests prior to writing code, ensuring tests are ready and in the pipeline. Traditional functional testing occurs to fill in the gaps which automated testing does not cover. Validation efforts complement verification by demonstrating that the integrated system fulfills its intended purpose and operates effectively in realistic conditions. This is achieved through simulation environments, digital twins, and user-in-the-loop testing that replicate mission scenarios with high fidelity. Together, these combined verification and validation activities ensure that critical systems perform safely, securely, and predictably in both expected and unforeseen operational environments.

INCOSE advocates for early and continuous V&V planning within systems development. Within a DevSecOps pipeline, this translates to defining V&V objectives as part of each sprint, embedding them into automated build scripts, and maintaining configuration-controlled evidence repositories. Every successful software build and deployment becomes a verified and validated event, complete with supporting documentation generated directly from the pipeline. This automation of assurance transforms V&V from a discrete phase into a living process that evolves in parallel with the system.

7.2.4 Security as a Systems Engineering Function

The "Sec" in DevSecOps introduces an explicit emphasis on integrating cybersecurity throughout the development lifecycle rather than treating it as an afterthought. Systems engineering naturally complements this integration by applying risk-based thinking, threat modeling, and architectural resilience analysis. In INCOSE terminology, this is the process of proactively identifying and mitigating design errors and vulnerabilities as early as possible in the system's lifecycle to reduce cost and complexity, which ensure that security, safety, and reliability are co-engineered rather than sequentially imposed.

In practical terms, systems engineers collaborate with developers to ensure that security testing is embedded into the CI/CD pipeline. This includes static code analysis, vulnerability scanning, and penetration testing. Each deployment undergoes automated validation against defined security baselines and configuration controls. When combined with architectural modeling and risk analysis, these measures establish a continuous feedback loop between development, operations, and assurance functions. This approach not only hardens the system against attack but also creates a verifiable chain of evidence for accreditation and certification authorities.

7.2.5 Configuration Management and Traceability in CI/CD

One of the central challenges in applying DevSecOps to critical systems is maintaining configuration integrity and full traceability across a stream of continuous builds and deployments. Without strict control, small configuration errors can propagate rapidly and compromise system performance, security, or mission readiness. Systems engineering principles address this challenge through rigorous configuration management practices that define how every element of the system is controlled, recorded, and verified. Each build artifact, source code version, model revision, and deployment image must be uniquely identifiable and consistently linked to a documented baseline of requirements, design data, and verification evidence. By maintaining this level of discipline, configuration management ensures that every delivered product can be reconstructed, audited, and trusted throughout the entire system's lifecycle.

Within a CI/CD pipeline, systems engineering principles manifest using controlled repositories, automated tagging, and immutable artifact storage that together maintain configuration integrity. These repositories serve as the authoritative record of every change, ensuring that source code, models, and configurations are versioned, traceable, and fully recoverable. Change control processes that were once managed manually through configuration review boards are now implemented as automated approval gates within the pipeline, ensuring that the rigor of governance is preserved even in an automated environment. Each approval gate evaluates compliance with established technical and programmatic baselines before any code or configuration is promoted from development into operational use. This automation does not replace engineering judgment; rather, it enforces consistent oversight by embedding policy, traceability, and validation rules directly into the delivery process. The result is a digital configuration management system that enables every delivered component to be reconstructed, audited, and verified against its defined baseline. Such rigor is essential for mission assurance and is a fundamental requirement for any software-intensive critical system where reliability, accountability, and safety must be proven at every stage of delivery.

7.2.6 Cultural Integration: Rigor and Agility in Balance

The successful application of systems engineering principles in DevSecOps is as cultural as it is technical, requiring a fundamental shift in how both developers and project managers approach their work. Software developers have traditionally operated with high autonomy and a focus on rapid iteration. In an environment of integrated SE and DevSecOps, they must embrace systems thinking and accountability for the broader mission context of their software. They need to understand not only how their software functions but also how it interacts with other components, impacts performance, and contributes to overall system assurance, as well as the real-world risks associated with failure. Project managers, on the other hand, must evolve from enforcing rigid schedules and documentation milestones to facilitating environments that support continuous verification, collaboration, and adaptive planning. This requires fostering a trust relationship between the development team and the project leadership.

Within DevSecOps, this cultural realignment means building a shared understanding that governance is a mechanism that ensures safe and reliable delivery and not a barrier to "slow things down" and prevent a successful release. Regular technical reviews, automated compliance checks, and digital traceability replace static document reviews and manual audits, allowing governance to occur continuously within the workflow rather than after the fact. This transformation preserves agility while embedding accountability into every stage of the development pipeline. When developers view systems engineering artifacts as living tools instead of bureaucratic obligations, and when project managers value rapid, evidence-driven insight over static reports, both groups contribute to a more resilient and transparent engineering culture. Ultimately, when properly implemented, this convergence produces a self-regulating ecosystem in which speed and rigor coexist harmoniously, ensuring that every software release strengthens both mission capability and stakeholder confidence.

7.2.7 *Final Thoughts*

DevSecOps and continuous integration and continuous deployment have fundamentally transformed both the pace and structure of software development within modern engineering environments. These practices enable organizations to deliver software capabilities at unprecedented speed, often several times faster than traditional acquisition or development models. For commercial systems, this rapid cadence is an advantage that supports innovation and responsiveness to user demand. For critical systems, however, such as those supporting defense, aerospace, energy, or healthcare missions, velocity alone is not enough. The consequences of a defect or misconfiguration in these systems can be severe, leading to operational failure, mission loss, or even harm to human life. Therefore, while DevSecOps accelerates delivery, it must be balanced by the rigor and assurance that only systems engineering can provide.

Without the integrating discipline of systems engineering, DevSecOps and CI/CD risk prioritizing speed over stability and innovation over safety. Systems engineering ensures that rapid iteration does not result in fragmented architecture, uncontrolled configurations, or unverified dependencies. It introduces traceability and accountability into what could otherwise become a continuous stream of untested code movement. Within this framework, DevSecOps becomes a continuously verified and validated system of systems, where every activity contributes measurable evidence of compliance, safety, and mission readiness.

The systems engineering approach begins by embedding V&V within every layer of the DevSecOps pipeline. Rather than treating testing as a final event, V&V activities occur continuously and automatically, ensuring that each code commit, model update, and configuration change is evaluated against defined acceptance criteria. Automated testing frameworks and digital engineering environments generate objective evidence, creating a living record of system quality throughout the lifecycle. This approach allows organizations to sustain both agility and rigor, demonstrating that rapid development cycles can coexist with the highest standards of reliability and safety. Verification ensures that every requirement is implemented correctly, while validation confirms that the system continues to meet its intended mission purpose. When applied together within an automated CI/CD pipeline, these activities create a self-reinforcing feedback loop that reduces defects, accelerates recovery from failure, and strengthens stakeholder confidence.

Equally vital is the integration of configuration management and security assurance into the DevSecOps environment. In a continuous delivery model, hundreds or even thousands of small changes can be introduced daily. Without systematic control, these changes can lead to divergence from requirements between deployed versions, loss of traceability, and compromised security. Systems engineering principles prevent this by ensuring that every artifact, whether source code, model, test script, or container image, is uniquely identifiable, traceable to a baseline, and subject to version control. Automated approval gates and digital configuration repositories replace manual audits, enforcing technical and programmatic compliance at every stage of delivery. When paired with embedded security checks, vulnerability scans, and compliance verifications, this digital configuration management system provides not only visibility but also assurance that every build is both technically sound and operationally safe.

System security directly benefits from this integration. In traditional development environments, cybersecurity considerations are often applied at the end of the lifecycle. In DevSecOps, security becomes intrinsic to every phase of development. Systems engineering ensures that these security mechanisms are not reactive but proactive, guided by architecture-level threat modeling like MBSEsec, risk assessments, and design-for-assurance principles. This approach transforms cybersecurity from an operational afterthought into a continuous engineering function. Every commit, build, and deployment is automatically scanned and validated against defined risk thresholds, creating a constant verification loop that protects both the integrity of the system and the confidentiality of its data.

Another benefit of applying systems engineering to DevSecOps lies in its governance model. Traditional systems engineering relies heavily on technical reviews, audits, and milestone-based evaluations. DevSecOps, by contrast, operates continuously, often blurring the boundaries between phases of development, test, and deployment. By incorporating systems engineering governance within the

DevSecOps pipeline, organizations can preserve the assurance benefits of structured reviews while eliminating the delays associated with manual oversight. Digital technical reviews, automated compliance reporting, and real-time performance dashboards allow program managers and chief engineers to monitor development progress and system health continuously. The same principles that guide formal system requirements reviews or test readiness reviews can now be implemented in real time, supported by data generated from the automated pipeline. This transformation maintains the rigor of systems engineering without slowing the cadence of delivery.

The integration of systems engineering within DevSecOps also requires cultural change across the organization. Developers must embrace the understanding that they are not simply writing code but building part of a mission-critical system whose success depends on compliance, traceability, and assurance. Program managers, meanwhile, must evolve from managing discrete tasks and deliverables to overseeing dynamic, continuously operating ecosystems of development, testing, and deployment. This alignment of mindset is essential, as the success of DevSecOps in critical systems depends as much on communication, transparency, and accountability as on technology.

By embedding systems engineering principles directly into DevSecOps workflows, organizations gain the ability to produce systems that are both fast and trustworthy. Each pipeline execution becomes a verification event, each deployment becomes a validated baseline, and each artifact carries traceable assurance evidence. This digital continuity eliminates many of the risks that have historically plagued rapid software development efforts. It allows programs to demonstrate compliance with standards such as INCOSE, ISO/IEC 15288, and DoD Digital Engineering Strategy objectives, while still achieving the responsiveness demanded by modern mission environments. The combination of continuous delivery with continuous assurance fundamentally redefines what it means to be agile in a high-stakes engineering domain.

In critical software systems, applying systems engineering principles to DevSecOps is not optional; it is the foundation of sustainable modernization. Systems engineering ensures that every element of the DevSecOps pipeline operates within a framework of accountability, verification, and mission alignment. When development, security, and operations converge under this unifying discipline, the result is a digital enterprise capable of delivering innovation rapidly while maintaining the safety, reliability, and trust that critical missions demand. This approach bridges the historical divide between agility and assurance, demonstrating that speed and rigor are not opposing forces but complementary strengths. By combining the structure of systems engineering with the adaptability of DevSecOps, organizations can achieve a state of continuous assurance, where every iteration strengthens system integrity rather than eroding it.

In the modern engineering landscape, the future of critical system development lies in this synthesis. DevSecOps provides the speed, automation, and scalability necessary to respond to evolving operational demands. Systems engineering provides the discipline, governance, and verification necessary to ensure that those responses are safe, effective, and sustainable. Together, they form a new paradigm for digital transformation in critical systems development where innovation and assurance advance together in continuous, measured harmony.

7.3 Discussion Questions

Verification and Validation Introduction: How does INCOSE's distinction between verification ("building the system right") and validation ("building the right system") influence engineering priorities for software-intensive critical systems, and what are the potential mission consequences of failing to integrate both perspectives early in development? In what ways does treating V&V as a continuous lifecycle function—rather than an end-of-project activity—change organizational accountability, cost structure, and defect-management strategies in critical software programs?

Verification: Building the System Right: Given INCOSE's emphasis on traceability, what practical techniques can ensure that every "shall" statement remains linked to its design implementation and verification evidence within a large, evolving system architecture? How do modern automation tools such as continuous integration pipelines, static analysis, and simulation frameworks enhance the objectivity and repeatability of verification in critical systems, and what new risks do they introduce if misapplied?

Goals of Verification Activities: Discuss the balance between functional and non-functional verification in high-reliability systems. Why is verifying attributes like safety, performance, and maintainability as essential as confirming basic functionality? Customer acceptance testing and user-in-the-loop evaluations are central to validation. What mechanisms can ensure that feedback from these activities meaningfully informs requirement updates without destabilizing configuration control?

Building the Right System: Validation focuses on stakeholder intent and operational fit. How can systems engineers reconcile technically verified software with user feedback indicating that the system still fails to meet mission needs?

Goals of Validation Activities: Customer acceptance testing and user-in-the-loop evaluations are central to validation. What mechanisms can ensure that feedback from these activities meaningfully informs requirement updates without destabilizing configuration control?

Integrated V&V Across the Lifecycle: What organizational and cultural barriers typically prevent continuous V&V adoption, and how can the principles of SEFAD or Agile systems engineering overcome resistance to embedding V&V in daily workflows? Explain how integrating verification and validation within CI/CD pipelines changes the traditional role of the systems engineer. Should engineers now act as curators of automated assurance evidence rather than gatekeepers of milestone reviews?

The Imperative of Rigor in V&V: INCOSE emphasizes independent review and objective evidence as cornerstones of credible V&V. How should organizations balance automation and human oversight to maintain both speed and integrity when verifying safety-critical or mission-critical systems?

Integrating the Three Ways of DevSecOps: How can the principles of flow, feedback, and continuous learning be practically engineered into a DevSecOps pipeline for a critical system where mission assurance cannot be compromised? Consider how these "three ways" influence the design of work handoffs, defect detection speed, and adaptation to operational feedback in a high-risk environment.

Systems Engineering Governance within DevSecOps: What specific governance mechanisms from INCOSE's Systems Engineering Handbook (such as technical reviews, requirement baselining, or verification traceability) can be embedded within DevSecOps pipelines to provide continuous oversight without slowing delivery? Discuss how these mechanisms might differ between a commercial software product and a defense or aerospace system.

Traceability and Digital Assurance in CI/CD: In a fully automated continuous integration and continuous deployment environment, what methods or digital engineering tools best ensure that every commit and configuration change remains traceable to verified requirements and validated mission objectives? How can teams create a "digital thread" that connects design intent, implementation, and operational evidence of assurance?

Balancing Speed with Safety through Embedded V&V: What are the major benefits and potential drawbacks of integrating verification, validation, and security assurance throughout the DevSecOps lifecycle for software-intensive critical systems? Discuss how early and continuous testing impacts system reliability, accreditation timelines, and risk posture compared to traditional end-of-cycle testing models.

Cultural Transformation toward Integrated Assurance: How must the culture of development and program management evolve to treat governance, security, and compliance as integral parts of delivery rather than procedural bottlenecks? What strategies can leadership apply to align engineers, developers, and program managers around a shared understanding that assurance and agility are mutually reinforcing goals?

Reference

1. International Council on Systems Engineering, *Systems Engineering Handbook*, 5th edn. (International Council on Systems Engineering (INCOSE), San Diego, 2023)

Chapter 8
Balancing Agility and Rigor

8.1 Why Agility Is Important in Software Development

Agility in software development is the disciplined ability to adapt, learn, and deliver value under changing conditions. It should not be a casual form of flexibility but rather focus on engineered responsiveness rooted in continuous feedback, collaboration, and iteration. Modern software systems rarely exist in isolation; they are part of intricate, interdependent ecosystems that evolve as quickly as the technologies that sustain them. In this environment, requirements are fluid, customer feedback drives innovation, and new vulnerabilities or opportunities emerge unexpectedly. Agility provides the structure and mindset required to navigate this constant state of change without sacrificing quality or control.

8.1.1 Responding to Change in Dynamic Environments

Software development operates within one of the fastest-changing technological domains in human history. Programming languages evolve, dependencies shift, user expectations grow, and operating environments diversify. What was state-of-the-art a year ago is probably obsolete today. Traditional development approaches, which assume stable requirements and predictable progression, struggle to keep pace with such volatility. Agility embraces change as inevitable, embedding responsiveness into the development process itself. Through iterative planning, continuous integration, and incremental delivery, Agile teams can respond to evolving needs while maintaining a steady rhythm of progress. This allows organizations to remain aligned with mission objectives, even as the surrounding context transforms.

In critical software systems, Agile provides an essential counterbalance to rigidity. While rigor ensures that verification, validation, and assurance activities are

D. Rosson, *Systems Engineering in Agile Software Development*,
https://doi.org/10.1007/978-3-032-14613-7_8

performed, agility ensures that these activities remain relevant in the face of new discoveries. When properly implemented, Agile principles prevent engineering from drifting into stagnation. Requirements are revisited rather than assumed correct; architectural choices are validated through prototypes rather than frozen by speculation. This cycle of learning and correction prevents the accumulation of hidden risks that later become the root cause of costly defects or program delays.

8.1.2 Learning as a Core Engineering Activity

Software development is framed in Agile as a continuous process of learning rather than a linear process of production. Each iteration, test event, or stakeholder review generates new information about the system's behavior, usability, and performance. Agile teams are structured to absorb this information rapidly. They interpret it collaboratively and rapidly, integrating it into their development methodology in attempt to improve design and planning. This feedback-rich environment reduces uncertainty and accelerates problem discovery.

In practice, this means that Agile teams do not wait until the end of a project to evaluate success. They test assumptions early and often, using working software as the primary vehicle for learning. In Scrum, this is at the Sprint Review. Continuous integration and automated testing ensure that defects are detected close to their source, when they are cheapest and easiest to fix. Retrospectives provide structured reflection, helping teams identify process inefficiencies and technical weaknesses before they become systemic. This cycle of inspection and adaptation is applied continuously and deliberately, ensuring a constant feedback learning loop in the development cycle.

8.1.3 Aligning Technology with Human Need

Agility ensures that software remains aligned with human need. Stakeholder engagement continues throughout the development cycle and is not limited to requirement definition at the project's outset. By maintaining direct and ongoing communication between users, developers, and engineers, Agile teams ensure that evolving expectations are captured and addressed in real time. This collaboration transforms software from a static product into a responsive capability that grows with its users. Backlogs are frequently reviewed and reprioritized as customer feedback is absorbed, and knowledge of the problem domain expands.

In critical systems, this alignment can be lifesaving. Consider applications involved critical software systems in which operational realities often differ from initial assumptions. An Agile approach allows developers to incorporate field feedback rapidly, updating logic, interfaces, or automation routines to reflect the truth of use rather than the theory of design. By shortening the feedback loop between the end user and the engineering team, agility ensures that systems remain effective, usable, and relevant throughout their operational life.

8.1.4 Managing Complexity Through Iteration

Modern software systems are rarely monolithic. They are composed of distributed services, interconnected APIs, third-party libraries, and complex security layers. This complexity cannot be fully anticipated during initial design. Agility offers a way to manage it iteratively, breaking large problems into smaller, testable increments that can be validated independently and integrated progressively. Each increment becomes an opportunity to test architectural soundness and to expose integration risks before they propagate.

By using incremental development, teams maintain a constant state of partial completion that is always functional and verifiable. This contrasts sharply with traditional "big bang" approaches, where integration occurs only after months or years of isolated component development. In Agile environments, integration is continuous, meaning that system behavior is constantly tested against reality. This approach not only reveals defects early but also builds confidence across teams and stakeholders who can see progress unfolding transparently.

8.1.5 Reducing Risk and Increasing Predictability

It may seem counterintuitive, but the proper application of agility can lead to increases in predictability. By delivering working software at regular intervals, teams provide stakeholders with tangible results and objective data about performance, schedule, and scope. This transparency replaces speculation with evidence. When issues arise, they are visible immediately, allowing for rapid decision-making and informed course correction. In contrast, traditional development methods often mask emerging risks until late in the process, when cost and complexity make them harder to address.

Agility also spreads risk across time rather than accumulating it at the end. Each iteration is a bounded experiment with measurable outcomes. Teams learn continuously about technical feasibility, user satisfaction, and system resilience. This distributed risk management approach transforms uncertainty into manageable increments. The result is not only more adaptive software but also more reliable project governance.

8.1.6 *Agility as an Organizational Capability*

Ultimately, agility is not confined to the software team; it is an organizational capability. It reflects how an enterprise learns, how it makes decisions, and how it adapts to new information. An organization that practices agility consistently across projects and departments develops an institutional reflex for learning. It becomes faster at identifying emerging challenges and more skilled at integrating solutions from diverse disciplines.

For engineering organizations, adaptability is not a convenience; it is a condition of survival. The pace of technological change has accelerated to the point where product cycles, once measured in years, now turn over in months. Markets evolve rapidly, new regulatory landscapes emerge, and mission contexts shift with geopolitical and environmental pressures. In such an environment, organizations that cannot adapt risk becoming experts in yesterday's problems. They may continue to deliver technically flawless systems that no longer serve the needs for which they were designed. Agility guards against this form of obsolescence by ensuring that innovation remains synchronized with relevance. It establishes a living connection between what an organization can build and what the world needs. This alignment allows teams to evolve architecture, update tools, and reimagine processes without losing continuity or purpose. In doing so, agility not only produces better software but also strengthens the organization itself, transforming it into a learning enterprise capable of sustaining performance, quality, and mission focus across generations of technology.

8.1.7 *Conclusion: Agility as the Rhythm of Modern Engineering*

Agility is the rhythm of modern software engineering. A pulse of learning, feedback, and adaptation that keeps technology aligned with an ever-changing world. It transforms development from a static act of construction into a living, evolving process of discovery. When practiced with discipline, agility does not undermine rigor; it enhances it by ensuring that verification and validation occur continuously, not episodically.

In the long view, agility is what allows engineering organizations to remain vital in a world defined by constant motion. Technology, markets, and missions do not stand still, and neither can the systems designed to serve them. Agility provides the mechanisms through which an organization senses change, interprets it intelligently, and responds without losing direction or integrity. It transforms the act of engineering from a static exercise in production into a dynamic practice of continual adaptation. In this sense, agility is not a temporary advantage; it is an enduring capability that determines whether an organization evolves or fades.

To view agility merely as a path to faster delivery is to misunderstand its purpose. Speed is a byproduct, not the goal. The true value of agility lies in its philosophy of alignment—maintaining coherence between design intent, user need, and operational reality even as all three evolve. It compels engineers to remain engaged with the systems they build, to listen to data from the field, and to adjust their assumptions considering evidence. Agility ensures that the feedback loops between creation and use remains open and healthy, enabling systems to grow in relevance rather than decay into irrelevance.

Rigor ensures that software works correctly; agility ensures that it continues to matter. Without rigor, systems fail. Without agility, they become obsolete. Both are essential, and both must coexist in balance for an engineering organization to thrive. Agility provides the motion that keeps rigor from calcifying into bureaucracy. Rigor provides the structure that keeps agility from dissolving into chaos. Together, they form the living equilibrium that defines resilient engineering practice.

Ultimately, agility is the quality that keeps engineering alive. It is the organization's capacity to learn faster than the environment changes, to sense shifts in context, and to act with purpose rather than panic. It preserves relevance, renews innovation, and sustains trust between engineers and the people they serve. In a discipline built on precision and predictability, agility is the reminder that excellence is not a destination, but a continuous act of adaptation.

8.2 Why Engineering Rigor is Important in Software Development

8.2.1 The Role of Engineering Rigor in Critical Software Systems

Engineering rigor is the disciplined use of structured reasoning and evidence to justify design decisions, test assumptions, and verify outcomes. It distinguishes confident engineering from hopeful guesswork by ensuring design elements are justified and validated, especially in software systems where failure may not appear until later stages. Rigor differentiates engineering from experimentation, turning discoveries into predictable performance. It converts innovative concepts into reliable and repeatable outcomes, preventing systems from being merely collections of ideas rather than dependable architectures.

For critical software systems like those in aircraft, medical devices, satellites, and finance, rigor is essential for trust. In these settings, even one unchecked assumption or error can lead to severe consequences, making thorough validation of every function vital. Neglecting rigor leads to latent defects that surface during rare but crucial events, causing failures when reliability is most needed. Such breakdowns stem from lapses in analysis and verification, undermining confidence in the system.

Ultimately, engineering rigor turns uncertainty into assurance. Far from hindering innovation, it safeguards and enables scalable creativity. As dependence on software grows, rigor becomes the foundation for adaptability and trust, ensuring systems do more than function but will perform capably and acceptably within the confines of the problem domain.

8.2.2 Defining Rigor in Context

Engineering rigor is often misconstrued as excessive bureaucracy, yet it does not equate to producing excessive documentation or following arbitrary procedures. Genuine rigor instills confidence: every requirement is comprehensively understood, each assumption systematically validated, interfaces are verified, and risks mitigated to acceptable levels. It assures predictable system behavior in both standard and exceptional scenarios. True engineering rigor ensures that capabilities required for proper execution of the system within the confines of the problem domain are understood and accounted for.

During the creation of critical systems, rigor provides traceability between required functionalities and implementation, traced through operational, logical, and physical views. It bridges design intent with operational outcomes and demands substantiation through evidence instead of intuition for claims regarding performance or safety. In the absence of engineering rigor, development teams risk losing crucial understanding of why a system works, increasing the chances that defects and currently unknown critical functionalities through a problem domain uncertainty and possible failure as conditions evolve.

8.2.3 Why Critical Systems Demand Rigor

Critical systems operate in demanding environments, often requiring decades-long functionality. They must embrace adaptation as technological shifts occur, disruptive technologies emerge, and sub-system and component obsolescence issues arise. During all of this, they must maintain uninterrupted availability under stress. In these settings, failure carries considerable human, financial, or strategic repercussions, making rigorous engineering and management essential.

Software compounds this challenge, as defects frequently remain undetected until they cause adverse effects. Unlike physical systems, software defects can often remain undetected until they manifest. Even minor oversights, such as an unchecked variable, can have far-reaching consequences within interconnected components. Rigor establishes quality and predictability at every phase, from requirements analysis to code reviews, configuration control, and verification testing, ensuring intended system behavior amidst complexity.

8.2.4 Rigor and the Engineering Mindset

Having good engineering and technical rigor are fundamentally a mindset, shaping how engineers approach problem-solving, evaluating trade-offs, and defining project completion. Rigorous professionals scrutinize assumptions, validate dependencies, and respond to ambiguity with further analysis rather than expedience. This attitude requires intentional cultivation, given persistent pressures such as schedules, resources, and unwarranted confidence that threaten its maintenance.

Rigor may erode over time as organizations prioritize expediency—documentation becomes perfunctory, review processes shorten, and verification steps are postponed. While these shortcuts seldom result in immediate failure, their cumulative effect fosters technical debt and organizational fragility, with latent faults jeopardizing safety margins.

8.2.5 Balancing Rigor and Agility

Rigor should not be conflated with inflexibility. The belief that disciplined engineering excludes adaptability is misguided; in fact, rigor facilitates agility by enabling controlled evolution of complex systems. Well-defined interfaces, traceable requirements, and repeatable verification processes empower teams to make confident adjustments. Whereas agility without rigor breeds disorder, rigor without agility induces stagnation. Effective engineering achieves both through structured flexibility rooted in robust principles.

8.2.6 The Cost of Abandoning Rigor

A lack of rigor typically escapes immediate detection, manifesting in overlooked edge cases, untested failovers, and undocumented assumptions. When failures occur, post-event analyses often reveal gaps: tests omitted, requirements misunderstood, or verification incomplete. Without rigor, system reliability remains provisional, susceptible to unexpected breakdowns.

Historical incidents highlight these risks: the Ariane 5 Flight 501 crash, Therac-25 radiation overdoses, and the 2024 CrowdStrike outage all resulted from defects being introduced to software systems resulting from a lapse in validation or verification. Each demonstrates how a stronger culture of engineering rigor, emphasizing traceability, peer review, and simulation, might have prevented substantial harm.

8.2.7 Rigor as a System Property

In systems engineering, attributes such as reliability, redundancy, and resilience are routinely modeled and measured. Rigor deserves similar treatment, as an emergent property stemming from people, processes, and artifacts alike. Through institutional mechanisms can be implemented and sustained, including configuration management, independent verification, test coverage metrics, and risk analysis.

Organizations that view engineering rigor as optional are, often without realizing it, designing for fragility. They may produce systems that function acceptably under controlled conditions, yet lack the resilience to withstand real-world complexity, change, or stress. In such environments, decisions are made reactively, verification is deferred, and assumptions accumulate faster than they are tested. The result is a brittle architecture—one that depends on good fortune rather than good engineering. By contrast, organizations that make rigor an integral part of their culture build a foundation for endurance. They recognize that disciplined verification, traceability, and continuous validation are not delays adversely affecting progress but rather result in sustained performance. These organizations view rigor as a form of risk management, quality assurance, and institutional memory, all operating in concert. The distinction between fragility and endurance may seem subtle in the short term, after all, both systems may appear to perform equally well at first. However, over time, it becomes decisive. One bends under pressure; the other adapts and continues to deliver.

8.2.8 Conclusion: Rigor as the Engine of Trust

Engineering rigor forms the cornerstone of trust within critical software systems. It converts creative concepts into dependable capabilities, ensuring that performance meets uncompromising standards where failure is not an option. Rigor supplies the necessary structure for innovation; guides change and substantiates assurance. It embodies both discipline and commitment and is a pledge that delivered systems will operate reliably and last.

Ultimately, rigor is not antagonistic to innovation; rather, it ensures that innovation endures. When performing engineering efforts, creativity without discipline produces novelty, but not reliability, and can directly lead to unstable systems. True innovation must persist beyond the prototype, operating under stress, at scale, and in unpredictable environments. Rigor provides the structure that allows bold ideas to mature into dependable systems that meet mission needs. Through careful verification and validation, requirements documentation, traceability, and modeling, rigor gives innovation the foundation to evolve safely and sustainably. Without it, progress becomes a cycle of rediscovery in which lessons are learned repeatedly, and failures reappear in new forms. Rigor does not restrain creativity; it preserves it by ensuring that new ideas have the stability needed to last.

8.3 The Art of Balance

Engineering, at its core, is a discipline of balance. It lives at the intersection of creativity and constraint, speed and certainty, innovation and assurance. The fusion of Agile methodologies with systems engineering rigor is one of the most significant balancing acts facing the modern engineering enterprise. This integration cannot be achieved by process alone—it must also account for the culture, values, and behaviors of the people who practice it.

8.3.1 Culture: The Hidden Architecture of Engineering

Every process improvement, however technical, is rooted in a belief system. Behind each methodology sits a community of professionals shaped by experience, training, and shared assumptions about what "good engineering" means. A product owner managing user stories and a systems engineer writing interface control documents may both be pursuing quality, yet they do so through different lenses. One prizes responsiveness and learning, the other values traceability and assurance.

When organizations attempt to merge these disciplines without understanding the values that drive them, friction emerges. Resistance is rarely about process—it's about identity. People defend what has kept their systems safe or their products successful. If these underlying beliefs are ignored, even the most elegant process framework will fail in practice.

Culture defines how engineers make choices under pressure. It shapes what they view as sufficient, how they interpret risk, and where they believe effort should be spent. Some engineers instinctively reach for a test bench, others for a spreadsheet or a prototype. Culture determines which of those impulses prevails when time is short and stakes are high. It's invisible on a process diagram, but it governs all real decision teams make.

8.3.2 The Two Worlds: Tradition and Agility

Traditional systems engineers often view Agile practices with deep skepticism. They see Sprint reviews, user stories, and continuous delivery pipelines, but worry that the underlying architecture will decay or that verification will be deferred too long. Their caution is not misplaced. Many have spent careers in domains where reliability is sacred—submarines, aircraft, satellites, medical devices. For them, failure is not an experiment; it's a breach of public trust. Agile, at first glance, can seem reckless.

Agile developers, on the other hand, sometimes perceive systems engineering as a bureaucratic labyrinth. They see documents frozen in early phases, change control boards moving at glacial speed, and heavy traceability requirements that appear disconnected from real customer needs. In their world, velocity and feedback are virtues. They measure success not in complete documentation but in value delivered to users. To them, systems engineering looks like a set of brakes installed on a race car built for responsiveness.

Both sides are partially right. Each perspective reflects hard-earned lessons from different operational environments. Systems engineers are right to insist on rigor in safety-critical contexts; Agile practitioners are right to demand adaptability when uncertainty dominates. The tension between the two is not a cultural one and not a technical contradiction.

8.3.3 The Cultural Divide

If we fail to confront this divide directly, integration efforts will falter. Organizations will adopt Agile ceremonies superficially while retaining the same rigid stage gates underneath, or they'll graft systems engineering documents onto Agile backlogs too late to matter. Process frameworks alone cannot close a cultural gap. Teams may comply outwardly yet revert to their defaults when deadlines tighten. Sustainable fusion requires a shift in mindset, not merely a change in method.

8.3.4 Bridging the Gap

So, how do we bridge this divide between rigor and agility? The first step is empathy. Leaders must create environments where systems engineers and Agile developers can speak openly, without fear of judgment, about what slows them down and what keeps them up at night. A Sprint retrospective with cross-disciplinary participation can reveal more about team dynamics than any formal review board. Similarly, pairing an Agile lead with a systems engineer for a project phase fosters mutual understanding through experience, not theory.

We must also clarify language. Words like traceability, verification, or interface control carry different meanings in different communities. Taking time to define these terms together—through workshops, glossaries, or shared templates—reduces confusion and builds shared ownership. When everyone understands why these artifacts exist and how they protect the mission, compliance becomes collaboration.

8.3.5 *Redefining Rigor*

It's time to redefine what we mean by rigor. Rigor is not about the number of documents produced or signatures collected. It is about confidence—confidence that the system will behave predictably under stress, that assumptions have been validated, and that interfaces will hold when reality tests them. Agile can be rigorous when it is built on solid architecture, disciplined testing, and controlled change. Systems engineering can be agile when it allows iteration, embraces learning, and evolves gracefully with discovery.

Rigor and agility are not opposites. They are orthogonal qualities—each reinforcing the other when applied thoughtfully. A system designed with both is not only faster to deliver but stronger in the face of uncertainty.

8.3.6 *Reinforcing the New Culture*

To make this balance sustainable, organizations must align incentives with desired behavior. If developers are rewarded solely for delivery speed, they will naturally sacrifice documentation and testing. If engineers are judged only by thoroughness, they may prioritize paperwork over progress. The solution is to reward integration: celebrate teams that deliver quickly and safely, that reduce cycle time while improving reliability.

Metrics should reflect both dimensions, velocity and assurance, adaptability and discipline. When leadership signals that both outcomes matter, culture begins to change. Within this new culture, we must identify and nurture what I call cultural translators. These are the hybrid professionals fluent in both dialects. They can facilitate backlog refinement in the morning and lead a hazard analysis in the afternoon. They are the connective tissue of transformation—the people who make integration real. Developing and empowering them is not optional; it is essential.

8.3.7 *SEFAD as an Example*

The Systems Engineering Focused Agile Development (SEFAD) model exemplifies the disciplined synergy between systems engineering's analytical framework and Agile development's adaptive methodology. Rather than prioritizing one approach, SEFAD demonstrates their coexistence throughout the lifecycle, resulting in mutual reinforcement. This model acknowledges contemporary engineering demands for rapid delivery, adaptability to uncertainty, and consistent excellence in safety, reliability, and mission assurance. SEFAD's principal value lies in sustaining the accountability and traceability of systems engineering while facilitating the speed and iterative feedback intrinsic to Agile practices.

Fundamentally, SEFAD operates on the premise that system-level intent should direct all iterative activities. Systems engineering establishes the architecture, interfaces, and constraints foundational to a program, while Agile enables ongoing architectural evolution via prototyping, incremental validation, and continuous learning. SEFAD ensures these iterations remain coordinated contributions toward a validated whole. This integration allows teams to deliver earlier, learn continuously, and adjust direction without compromising essential systemic integrity. The model replaces the perceived conflict between rapid delivery and structural stability with a paradigm of disciplined adaptability.

From a systems engineering standpoint, SEFAD maintains the rigorous analysis necessary for complex, multi-domain projects. Requirements are continually decomposed, traced, and validated, rather than being static artifacts. System architecture is modeled dynamically, allowing real-time visualization of interdependencies and change impacts. Verification and validation activities occur throughout the development lifecycle, ensuring compliance, performance, and safety mature alongside development. SEFAD effectively evolves the conventional "V-model" into an adaptive framework, matching Agile's pace while preserving structural coherence.

Agile teams benefit from SEFAD's structured environment, which transforms experimentation into evidence-based advancement. Operating with shifting requirements and uncertain conditions, Agile decisions are anchored within a traceable system model. Each iteration aligns with validated objectives and controlled interfaces; storyboards, Sprints, and backlogs are explicitly linked to formal requirements and architecture elements, establishing clear connections to broader mission outcomes. This structure mitigates the risk of divergence from intended design, or "Agile drift," enhancing both flexibility and discipline.

SEFAD incorporates comprehensive engineering feedback loops across organizational levels. Test results, modeling data, and stakeholder input within each iteration are valued as enduring engineering evidence, informing updates to system models and baselines. This cycle fosters a continuously improving engineering ecosystem where design, analysis, and verification interact in near real time, promoting superior technical quality and strengthening stakeholder confidence, while minimizing late-stage rework.

Operationally, SEFAD brings notable improvements: reduced verification cycle times, fewer integration issues, and enhanced communication among cross-functional teams. By embedding traceability within Agile workflows, SEFAD prevents critical requirements or interfaces from being overlooked under demanding schedules. Teams can demonstrate incremental compliance, providing early assurance to auditors and program sponsors. For regulated industries such as defense, aerospace, and healthcare, SEFAD offers transformative capabilities for meeting certification needs without reverting to rigid legacy processes.

Culturally, SEFAD bridges the gap between traditional engineers and Agile practitioners, acknowledging essential strengths and fostering adaptation for success. Systems engineers adopt incremental discovery, accepting evolving requirements; Agile developers recognize the importance of traceability and controlled baselines for innovation sustainability. SEFAD introduces shared artifacts, metrics, and ceremonies that establish a unified culture of rigor and responsiveness.

The practicality of SEFAD is evident from its implementation in defense programs, digital twin projects, and enterprise transformations, addressing limitations of both traditional and pure Agile methodologies. These applications report less rework, improved verification, greater stakeholder satisfaction, and stronger mission alignment. SEFAD integrates existing frameworks, including Agile Scrum, Model-Based Systems Engineering (MBSE), and standard governance, into a cohesive operational philosophy.

Ultimately, SEFAD redefines engineering leadership by positioning rigor and agility as complementary competencies. It encourages leaders to think holistically, treat architecture as dynamic, and measure progress by verified capability in addition to velocity. SEFAD confirms that sustainable engineering requires both speed and structure, not exclusively one or the other.

Consider SEFAD as not solely a process but also a mindset. It instills intentional design, evidence-driven adaptation, and confident delivery. The model upholds engineering accountability while embracing modern software adaptability. For organizations managing complex, software-intensive missions, SEFAD serves as a strategic blueprint for success, illustrating that mastery of balance, rather than speed or strict adherence to procedure, defines the future of engineering.

8.3.8 Toward an Integrated Future

Ultimately, this transformation is not about choosing sides. It's about forging a culture that draws strength from both worlds. Systems engineering provides the structure; Agile provides the motion. One without the other is incomplete. Together, they enable engineering organizations to deliver systems that are robust, resilient, and responsive to change. This is a redefinition of what modern software systems engineering looks like. It requires mutual respect, shared language, and a collective commitment to delivering outcomes that matter.

If you work in critical software systems, whether defense, aerospace, healthcare, or finance, you no longer have the luxury of choosing between speed and structure. You must demand both. The SEFAD approach offers a disciplined way to achieve that balance, proving that with the right culture and rigor, engineering teams can move fast without breaking what truly matters.

8.4 Discussion Questions

Defining agility and rigor: How do agility and rigor differ in purpose yet complement one another in modern software development? Why does an engineering organization need both to remain effective?

Managing change: In what ways does agility enable organizations to respond to evolving requirements, technologies, and stakeholder needs without losing quality or control?

Rigor in critical systems: Why is engineering rigor particularly vital in safety-critical or mission-critical software systems, and what are the risks when it is neglected?

Learning and feedback: Agile emphasizes continuous learning and iteration. How can this feedback-driven approach improve technical quality and reduce long-term risk?

Cultural integration: The text describes culture as the "hidden architecture of engineering." How do cultural beliefs and professional identities shape the success or failure of Agile–Systems Engineering integration?

Balancing flexibility and discipline: How can teams maintain adaptability while preserving traceability, verification, and system assurance? What tools or practices support this balance?

Risk and predictability: Agility is often perceived as introducing uncertainty. How can disciplined Agile practices actually increase predictability and strengthen governance?

Innovation and sustainability: The chapter argues that "rigor ensures software works correctly; agility ensures it continues to matter." Discuss how this relationship sustains innovation over time.

SEFAD in practice: What key principles of the Systems Engineering Focused Agile Development (SEFAD) model illustrate how rigor and agility can coexist in the same lifecycle? How might these principles apply in your domain?

Leadership and the future of engineering: What new leadership mindset is required to balance agility and rigor in software-intensive systems? How should future engineering leaders measure success?

Index

D. Rosson, *Systems Engineering in Agile Software Development*, https://doi.org/10.1007/978-3-032-14613-7

MIX
Papier aus verantwortungsvollen Quellen
Paper from responsible sources
FSC® C105338

If you have any concerns about our products,
you can contact us on
ProductSafety@springernature.com

In case Publisher is established outside the EU,
the EU authorized representative is:
Springer Nature Customer Service Center GmbH
Europaplatz 3, 69115 Heidelberg, Germany

Printed by Libri Plureos GmbH
in Hamburg, Germany